TARGETED CANCER IMAGING

TARGETED CANCER IMAGING

Design and Synthesis of Nanoplatforms based on Tumor Biology

MEHDI AZIZI

HADI KOKABI

HASSAN DIANAT-MOGHADAM

MOHAMMAD MEHRMOHAMMADI

ELSEVIER

ACADEMIC PRESS

An imprint of Elsevier

Academic Press is an imprint of Elsevier
125 London Wall, London EC2Y 5AS, United Kingdom
525 B Street, Suite 1650, San Diego, CA 92101, United States
50 Hampshire Street, 5th Floor, Cambridge, MA 02139, United States
The Boulevard, Langford Lane, Kidlington, Oxford OX5 1GB, United Kingdom

Notices
Knowledge and best practice in this field are constantly changing. As new research and experience broaden our understanding, changes in research methods, professional practices, or medical treatment may become necessary.

Practitioners and researchers must always rely on their own experience and knowledge in evaluating and using any information, methods, compounds, or experiments described herein. In using such information or methods they should be mindful of their own safety and the safety of others, including parties for whom they have a professional responsibility.

To the fullest extent of the law, neither the Publisher nor the authors, contributors, or editors, assume any liability for any injury and/or damage to persons or property as a matter of products liability, negligence or otherwise, or from any use or operation of any methods, products, instructions, or ideas contained in the material herein.

Library of Congress Cataloging-in-Publication Data
A catalog record for this book is available from the Library of Congress

British Library Cataloguing-in-Publication Data
A catalogue record for this book is available from the British Library

ISBN: 978-0-12-824513-2

For information on all Academic Press publications visit our
website at https://www.elsevier.com/books-and-journals

Publisher: Mara Conner
Acquisitions Editor: Tim Pitts
Editorial Project Manager: Isabella C. Silva
Production Project Manager: Prasanna Kalyanaraman
Cover Designer: Greg Harris

Typeset by TNQ Technologies

Contents

Preface

Cancer is a complex disorder resulting from several alterations in biological processes and signaling pathways. Malignant tumors continue to be a source of high morbidity and mortality rates throughout the world. Cellular and molecular heterogeneity within a single tumor plays a key role in tumor progression and the failure of conventional therapies. Meanwhile, researchers are just beginning to understand the full panorama of cancer biology and whether cancer heterogeneity plays a role in specific cancer types. Before targeted delivery approaches can be rationally designed, a better understanding of the biological processes involved in the biodistribution, release, and retention of targeted delivery systems inside the tumor is imperative. Along with advances in cancer biology, new challenges and possible treatment options are at the forefront of cancer research. Complexity, heterogeneity, low concentrations of biomarkers, and unpredictable cancer cell behavior remain major barriers to developing novel treatment strategies.

Currently, two major strategies exist for targeting cancer: (1) Passive targeting exploits the accumulation of nanocarriers at the site of interest, such as tumors. The convection (or passive-diffusion process) is mediated by the transportation of nanocarriers via pores in leaky capillaries that are present in tumor masses and in tissues that trigger angiogenesis. This process occurs in conjunction with the enhanced permeability and retention (EPR) effect. However, passive targeting cannot be classified as a type of selective targeting. The EPR effect applies not only to tumors but also to off-target organs such as the spleen, liver, and lungs. (2) Active targeting agents can selectively transport nanoparticles into the tumor mass, and they bind to molecules expressed on the cancer cell surface with high affinity, leading to endocytosis-mediated cell uptake. A surface-functionalized nanocarrier using specific ligands can complement the passive targeting approach to improve nanocarrier delivery and tumor localization efficiency.

Although considerable efforts have been conducted to diagnose, improve, and treat cancer over the past few decades, existing therapeutic options are insufficient, as mortality and morbidity rates remain high. Perhaps the best hope for substantial improvement lies in early detection. Recent advances in nanotechnology are expected to increase our current understanding of tumor biology and allow nanomaterials to be used for targeting and imaging in both in vitro and in vivo experimental models. Nanostructures have intrinsic physicochemical characteristics that make them valuable tools, and nanostructures have received much attention within nanoimaging. Consequently, rationally designed nanostructures have been successfully employed in cancer imaging to target cancer-specific and cancer-associated molecules and pathways. This review categorizes imaging and targeting approaches according to cancer type and highlights some new safe approaches involving membrane-coated nanoparticles, tumor cell-derived extracellular vesicles, circulating tumor cells, cell-free DNA, and cancer stem cells in hopes of developing more precise targeting and multifunctional nanotechnology-based imaging probes in the future.

Regarding these limitations, nanotechnology is a promising field at the forefront of cancer detection research. Considerable effort has been made to create many different targeted molecular imaging nanoplatforms with unique features and capabilities. Unlike conventional imaging techniques, tumor-selective imaging probes could deliver an optimized imaging agent to a specific target with high affinity, specificity, and sensitivity. Secondly, lower but effective dosages of tumor-selective therapeutic nanoplatforms could efficiently localize in tumors with minimized systemic toxicity. Moreover, it is possible to monitor and confirm whether nanoplatform-based tumor-selective imaging probes have been properly delivered to the targeted site after injection. Compared with traditional imaging agents, the number of injected tumor-selective imaging probes that reach tumor sites can be quantitatively analyzed. Further, the biodistribution of these probes within the body can be monitored over long periods. Thus, targeted nanoplatforms have good potential to significantly increase imaging contrast, enable cancer detection at earlier stages, and provide monitoring of responses to conventional tumor therapy or molecular targeted therapy. Another benefit of using targeted nanoprobes is that imaging nanoplatforms can monitor changes in the molecular microenvironments associated with tumors. In addition, integrated imaging and therapeutic capabilities provide a combined diagnostic therapy known as a theranostic approach. Theranostic systems can reduce toxicity, enhance selectivity and targeting, generate data for diagnostics, and enhance therapeutic efficiency. The final goal of this book is to provide us with knowledge about the design and manufacturing of suitable targeted nanostructures for selective and specific targeting of tumor masses in the body.

CHAPTER 1

Introduction to cancer biology

1. Tumor biology

Cancer is a heterogeneous multifactorial disease and a primary cause of death worldwide. As with normal conditions, various codependent networks, including cellular, tissular, and organismic layers, are involved in the cancerization of normal tissues and even human bodies (systemic disease). Geographically, malignant tumors are formed by three different cancer tissues: the primary tumor, local metastasis, and distant metastasis [1]. Considering a Darwinian view of cancer development, Peter Nowell [2] proposed that colony formation of cancer cells occurred through gradual selection and expansion of the advantageous properties of malignant cells with specific, heritable genetic aberrations. The genome, somatic mutations, and epigenetic modifications are the focus of investigations of the cancer process. In addition, identifying nonsilent somatic mutations in phenotypically normal cells and participating stromal cells, immune cells, and tissue-relevant processes (e.g., angiogenesis) provides ideas for considering a broader study of cancer at tissue and organismic levels [1].

Historically, the hallmarks of cancer were presented in several studies by Hanahan and Weinberg [3], who organized cancer biology into six major hallmarks: self-sufficiency in growth signals, insensitivity to antigrowth signals, evasion of apoptosis, limitless replicative potential, sustained angiogenesis, and tissue invasion and metastasis. Over time, Hanahan and Weinberg [4] updated their views, and 6 years later, Fouad and Aanei [5] updated the picture of cancer hallmarks and presented other cancer-associated hallmarks, such as altered stress response favoring overall survival, the metabolic rewiring and abetting microenvironment, and immune modulation. According to evolutionary mechanisms, Alkhazraji et al. [6] recently re-presented promising hallmarks such as extensive genomic and epigenetic diversification, resistance to cell death, modulation of the modulatory microenvironment, and cellular plasticity. In this chapter, we present tumor biology that covered all the hallmarks studied to date.

Targeted Cancer Imaging
ISBN 978-0-12-824513-2
https://doi.org/10.1016/B978-0-12-824513-2.00002-4

Additionally, we describe novel tumor-associated/derived concepts and components that contribute to validating the hallmarks of tumor biology.

1.1 Cancer hallmarks

1.1.1 Constitutive proliferation

Cell proliferation refers to increases in cell number due to cell growth and division. The proliferation and growth of normal cells are regulated by releasing growth-promoting signals in a controlled manner to provide homeostasis under normal conditions. However, deregulating these signals in cancer cells gives them a survival edge and ensures higher proliferative capacity. The sustained proliferation of cancer cells occurs by binding growth factors/ligands to their cell-surface receptors, which contain intracellular tyrosine kinase domains to promote intracellular signaling pathways that induce cell cycle progression and cell growth. Cancer cell-derived growth factor ligands also bind to the receptors themselves (i.e., autocrine signaling) or to tumor-associated stroma (i.e., paracrine signaling) to sustain growth-promoting signal activation [4].

Elevated frequencies of DNA replication-associated errors and distinct somatic/chromosomal mutation signatures in cancerous cells result in constitutive receptor activation and overexpression, aberrant signaling, and impairment of receptor degradation machinery [5]. The deletion and point mutations that cause *EGFR* (epidermal growth factor receptor) amplification and overexpression cause rapid growth, metastasis, and invasion in highly aggressive lung tumors [7]. Mutationally activated *B-Raf* due to substitution of glutamic acid for valine at amino acid 600 (V^{600E}) impairs B-Raf protein function, thus resulting in excessive activation of the downstream proliferative mitogen-activated protein kinase pathway in melanoma, glioblastoma, and thyroid, lung, and colon cancers [8]. Additionally, defects in negative-feedback mechanism-driven homeostatic regulation due to such mutations in *ras* gene such as oncogenic genes lead to hyper-Ras GTPase activity and enhance proliferative signaling [9]. Promoter methylation in PTEN is considered a loss-of-function mutation that causes amplified phosphoinositide 3-kinase (PI3K) signaling and promotes tumorigenesis in human cancers [10]. Moreover, activated PI3K induces ATP production by stimulating the mechanistic target of rapamycin, a coordinator of cell growth and metabolism that supports cell proliferation. Together, the promotion of cell division rates and inhibition of cell cycle arrest or apoptosis ensures unchecked cancer cell growth and tumorigenesis.

1.1.2 Insensitivity to antigrowth signals

Besides sensitivity to growth-stimulatory signals, cancer cells evade the effects of tumor suppressors that otherwise limit uncontrolled cell proliferation through transducing checkpoint or growth-inhibitory signals or by activating senescence and apoptotic programs. The loss-of-function mutation in inhibitory effectors such as retinoblastoma allows cancer cells to progress to replication with further newly acquired DNA aberrations as well as genomic instability and tumor heterogeneity [11]. Apoptosis inducers such as TP53 protein can halt further cell cycle progression under intracellular stress conditions (e.g., insufficient nucleotide pools, glucose, or oxygenation) or trigger apoptosis while facing irreparable damage of the cellular subsystem (e.g., after chemical exposure, UV irradiation, or oncogene activation) [12]. Therefore, any mutation in p53 gene encoding leads to uncontrolled division and accelerated proliferation.

Once cancer cells proliferate, highly dense populations of cells further reduce cell proliferation due to "contact inhibition." However, this homeostatic mechanism is abrogated during tumorigenesis. *NF2* gene coding merlin protein, as a tumor suppressor, provides contact inhibition through coupling cell surface-adhesion molecules (e.g., E-cadherin) to transmembrane receptor tyrosine kinases (e.g., the epidermal growth factor receptor) [13]. The loss-of-function mutation in the *NF2* gene leads to loss of contact inhibition and triggers a form of human neurofibromatosis [14]. Additionally, LKB1 epithelial polarity protein, as a tumor suppressor, maintains tissue integrity, and its downregulation causes Myc oncogene-induced transformation [15].

1.1.3 Resistance to cell death mechanisms

Usually, cell death through any natural mechanism such as necrosis, apoptosis, necroptosis, and autophagy should work as a barrier to cancer development. Yet pathways for cell death are attenuated or abrogated in those tumors that succeed in progressing to states of high-grade malignancy and resistance to therapy. Apoptosis or programmed cell death is characterized by the activation of caspases, which cleave cellular substrates, and through multistep processes, the dying cells are broken into apoptotic bodies that are taken up by phagocytic cells without triggering inflammation [16]. Tumors evade apoptosis through overexpression of antiapoptotic regulators (i.e., Bcl-2, Bcl-xL) or downregulating proapoptotic factors (i.e., Bax, Bim, Puma) and caspase proteins. Bcl-2 mediates the apoptotic signals from the upstream regulators and downstream effector components that the chromosomal translocation t [8,14] driving Bcl2 overexpression was cell documented and observed in follicular lymphoma. TP53 tumor suppressor function in response to irreparable DNA damage induces apoptosis that causes loss of TP53 or perturbates the apoptosis-inducing circuitry [4].

Autophagy begins with the engulfment of cytoplasmic materials in the autophagosome, followed by fusion with lysosomes to form the autolysosome. Thereafter, autolysosome-associated catalytic enzymes degrade the cargoes, and by-products are recycled into the cytosol to maintain cellular homeostasis and survival [17]. The intrinsic stress such as hypoxia, nutrient deprivation, pH changes, and extrinsic insults such as chemotherapy and radiotherapy act as inducers of autophagy in cancer cells, further assuring tumor cell survival, treatment resistance, and cancer relapse [18]. On the other hand, the autophagy machinery is linked to apoptosis, and inhibitors of apoptosis similarly inhibit autophagy and thus ensure cancer cell survival. The deletion or loss of beclin (encoding an essential autophagy protein) was indicated in breast, ovarian, and prostate cancer cases and attributed to the role of autophagy in eliminating damaged mitochondria during periods of stress, reducing the burden of reactive oxygen species (ROS) [17].

Apoptotic cells are progressively disassembled and then consumed by its by phagocytic cells. In contrast, necrosis is thought to be independent of the activity of caspase. In the end, necrotic cells become bloated and explode and release their contents, such as proinflammatory signals, which are followed by recruiting the inflammatory cells of the immune system into the surrounding tissue microenvironment. In neoplasia, inflammatory immune cells contribute to cancer progression by promoting angiogenesis, cancer cell proliferation, and invasiveness. Necrotic cell-derived high mobility group 1 (HMGB1) protein and interleukin (IL)-1α induce tumor-promoting inflammation and proliferation of tumor-supportive cells, respectively [19,20]. Moreover, necrotic tumor cell-released potassium suppresses the antitumor immunity CD4 and CD8 T cells [16].

Necroptosis and ferroptosis are two types of regulated necrosis that act as alternative ways to eradicate apoptosis-resistant cancer cells. When apoptosis is blocked, tumor necrosis factor triggers certain cells to undergo regulated necrotic cell death by a caspase-independent process termed necroptosis [16]. One study reported radiation-induced lung tumor cell necroptosis followed by the release of immune-activating HMGB1 from necroptotic cells [21]. While promising, it seems that necroptosis possesses both tumor-suppressive and tumor-supportive effects, and similar to necrosis, necroptotic cells trigger inflammatory responses. Thus, induction of necroptosis for cancer therapy elevates antitumor immunity responses [16]. Finally, ferroptosis is an iron-dependent form of cell death and occurs through excessive peroxidation of polyunsaturated fatty acid-containing phospholipids in mammalian cell membranes [22]. Therefore, ferroptosis inducers create high expectations for potential cancer therapy strategies based on ferroptosis induction. Ferroptosis is regulated by p53, and thus, any dysregulation in this cooperation gives cancer cells a survival benefit under metabolic stress due to lipid oxidation [22].

1.1.4 Evading replicative-induced mortality

Cancerous cells escape from limited proliferation elucidated by senescence and crisis/apoptosis. Senescence is defined as the process of irreversible exit from the cell cycle and occurs against cellular stress such as nontelomeric DNA damage (e.g., ROS accumulation) and telomeric DNA damage due to shortening of telomeres after exhaustion of replication potential. Cancer cells overwhelm telomere erosion through overexpression or reexpression of the components of the telomerase complex that are capable of protecting and reconstituting the telomeric ends and maintaining sufficient length for further replication [23]. Telomerase is a specialized DNA polymerase that adds telomere repeat segments to the ends of telomeric DNA absent in nonimmortalized cells but significantly reexpressed in cancer cells as spontaneously immortalized cells. Therefore, telomerase activity resists cancer cells against proliferative barriers, including senescence and apoptosis. Moreover, the noncanonical role of telomerase is relevant to its protein subunit, named telomerase reverse transcriptase (TERT) that amplifies proliferative-inducer signaling pathways (e.g., WNT and MYC pathways) [24] and induces activation of the antiapoptotic NF-kB signaling pathway [25]. TERT also suppresses antiproliferative signaling induced by the TGF-β pathway [23] and provides stemness and metastasis through induction of epithelial–mesenchymal transition (EMT)-relevant markers in stemlike cancer cells [26]. Overall, telomerase plays a protective role in the genetic integrity and propagation of a given advantageous cancer cell clone (Fig. 1.1).

1.1.5 Inducing vascularization

Angiogenesis is a biological process through which new blood vessels are shaped from preexisting vasculature. Like normal tissue, tumorigenesis demands the sustained ingrowth of blood vessels to provide oxygen and nutrients and to remove waste products. Angiogenesis occurs through several major, sequential steps: (1) proteolytic enzyme-induced degradation

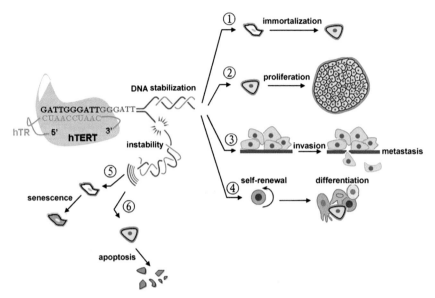

Figure 1.1 Reexpression of TERT provides cancer cell survival through several pathways including (1) immortalizing primary human cells; (2) insuring tumor cell proliferation; (3) inducing invasion and metastasis; and (4) providing stemness and pluripotency; targeting TERT results in telomere loss and (5) nonprevention of stress-induced senescence in normal cells; and (6) tumor cell apoptosis. *(Copy from Lü M-H, Liao Z-L, Zhao X-Y, Fan Y-H, Lin X-L, Fang D-C, et al. hTERT-based therapy: a universal anticancer approach. Oncol Rep 2012;28(6):1945−1952).*

of extracellular matrix (ECM) components surrounding the blood vessels; (2) activation, migration, and proliferation of endothelial cell (ECs); and (3) transformation of ECs into tube-like structures and formation of capillary tubes, followed by the formation of novel basement membranes [28].

Angiogenesis demands the cooperation of several cell types, soluble angiogenic factors, and ECM components and is regulated by pro- and antiangiogenic effectors. Once tumorigenesis occurs, proangiogenic signaling dominates antiangiogenic signaling; this change is termed the "angiogenic switch" [29]. At the molecular level, proangiogenic factors including vascular endothelial growth factor (VEGF), tip cell-produced Delta-like-4 ligand, platelet-derived growth factor-B (PDGF-B), fibroblast growth factor 2 (FGF-2), angiopoietins, matrix metalloproteinase (MMP), ephrins, apelin, and chemokines support angiogenic switching through the transformation of quiescent ECs into ECs proliferative toward new blood vessel formation. For example, stroma- and tumor cell-derived

VEGFs bind and activate VEGF receptor-2 on ECs, which induces EC proliferation and angiogenesis. Tumor cell-derived MMPs induce remodeling of the ECM to facilitate tumor-associated angiogenesis. Myeloid-derived suppressor cells secrete CCL2, CXCL8, CXCL2, and CXCL12 ligands, which bind to their receptors on tumor ECs and increase angiogenesis [18,30]. Tumor-associated fibroblasts and mesenchymal stem cells are present in secret tumor niche VEGF, MMPs, and CXCL12 that can more provide neovascularization [18]. The aforementioned proangiogenic factors are produced by immune cell lymphocytes, neutrophils, and macrophages in tumor niches and influence angiogenesis. Finally, hyperproliferation of tumor cells induces hypoxic conditions due to increased oxygen consumption. Responding to hypoxia, hypoxia-inducible factors (HIFs, HIF-α, and HIF-β) mediate the upregulation of genes encoding VEGF, FGF, and PDGF leading to rapid flipping of the angiogenic switch for promotion of neovascularization and angiogenesis [30].

1.1.6 Tumor invasion and metastasis

Tumor metastasis is a dynamic biophysical process triggered by the dissemination of invasive tumor cells with an unstable genome. They have undergone ECM remodeling, the downregulating of E-cadherin, upregulating of N-cadherin, and transient emerging of the EMT process (Fig. 1.2). Therefore, they lose their polarity and cell−cell/matrix adhesion [32]. Primary tumor cell-secreted MMPs and the proteolytic urokinase-type plasminogen activator system induce ECM remodeling to trigger invasion. Moreover, primary tumor-released extracellular vesicles such as exosomes carrying and displaying invasion-promoting factors that drive metastatic properties induce EMT and ECM remodeling and promote tumor cell resistance to apoptotic signals [33]. After that, invasive cells detach from the primary tumor site and enter the peripheral blood (or intravasation) as circulating tumor cells (CTCs) in both single and cluster forms to migrate toward distant sites to drive metastatic tumor formation [31]. Sometimes, CTC-derived VEGF, MMPs, and IL-8 mediate the extravasation and seeding of CTCs at primary tumors in a "tumor self-seeding" process [31,34,35].

Collective intravasation and migration of CTC clusters (groups of >2 CTCs) demand cell−cell junction formation that is supported by upregulation of stemness and EMT markers and tight junctional proteins such as claudin-11 in single CTCs [31,36]. Compared with single CTCs, heterogenous clusters indicate higher metastatic potential due to their

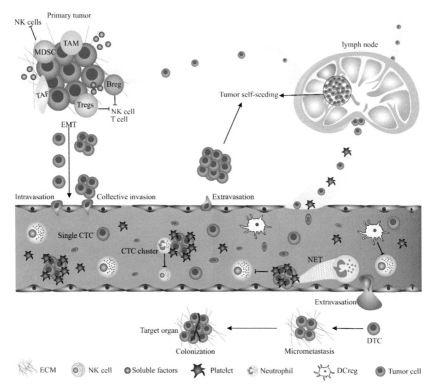

Figure 1.2 Metastasis and immune evasion of CTCs to induce tumor metastasis. Immune-resistant tumor cells detach from the primary tumor under the EMT process and ECM remodeling in both single and cluster forms. CTCs may reenter in local areas and lymph nodes to induce tumor self-seeding or pass intravasation to transfer to distant organs through the bloodstream after extravasation from micrometastasis and macrometastases. Immune-mobilized cells associated with CTC clusters provide protection against shear stress and immune cell cytotoxicity, and at distant sites, they facilitate CTC extraversion. Abbreviations: *Breg*, regulatory B cell; *DCreg*, regulatory dendritic cell; *DTC*, disseminated tumor cell; *ECM*, extracellular matrix; *EMT*, epithelial–mesenchymal transition; *MDSC*, myeloid-derived suppressor cell; *NET*, neutrophil extracellular trap; *TAF*, tumor-associated fibroblast; *TAM*, tumor-associated macrophage. *(Copy from Dianat-Moghadam H, Mahari A, Heidarifard M, Parnianfard N, Pourmousavi-Kh L, Rahbarghazi R, et al. NK cells-directed therapies target circulating tumor cells and metastasis. Cancer Lett 2021;497:41–53).*

colony-forming potential and higher survival against blood pressure and the immune system exerted by stromal and immune cells such as macrophages, fibroblasts, and neutrophils, and platelets [32] (Fig. 1.2).

Low shear stress and size limitations (CTCs of up to 20 μm in diameter vs. capillaries of ~3 to 7 μm), VEGF, exosomes, and MMPs all provide

CTC extravasation through EC and ECM remodeling at distant sites. The mesenchymal—epithelial transition process restores cellular traits of the primary tumor and provides collective migration of clusters to colonize at distant organs and create disseminated tumor cells (DTCs). Organotropism of DTCs must adapt to the premetastatic niche, where they are exposed to interactions with stroma cell types, resident niche cells, exosomes, ECM proteins, and deleterious signals [33]. Finally, DTCs may act as metastasis-initiating cells, generate micrometastatic lesions, and create minimal residual disease. Existing stressful stimuli and inter-DTC stemness markers may remain in dormant states for years or decades until reemerging to form macrometastases and induce metastasis [31,32].

1.1.7 Immunomodulation

The innate and adaptive immune system should eradicate or at least control tumor cell activity. The cross-talk between immune cells and tumor cells reaches a dynamic equilibrium to restrain tumor progression. However, suppressing antitumor immune responses and inducing an immune-suppressive tumor microenvironment (TME) leads to clonal selection and evasion of immune effector-resistant tumor cells [32]. $CD8^+$ cytotoxic T cell (CTL)- and $CD4^+$ helper T (Th)1 cell-produced interferon-γ/cytotoxins are expected to suppress cancer development; however, over-activation of immune cells causes inflammation and promotion of cancers such as those documented for the hepatitis B/C viral-induced inflammatory state and that appear in hepatocellular carcinoma [37]. Tumor-supportive immune cells such as regulatory T cells (Tregs), MDSCs, and M2 macrophages exert tumor immune escape by creating an inflammatory TME, suppressing CTL, and upregulating VEGF, TGF-β, and IL-10, all of which are mediators of antitumor-immune suppression, angiogenesis, and metastasis (Fig. 1.2). VEGF, IL-10, and TGF-β are known as inhibitors of the differentiation of progenitor cells into mature dendritic cells (DCs) [38]. In addition, TGF-β and IL-10 promote immune escape by inducing the differentiation of Th1 to Th2 (immune deviation) [39].

Mutational genetic changes or instability results in the production of heterogeneous tumor antigens that complicates their recognition and elimination by immune cells [40]. Moreover, cancer cells downregulate the major histocompatibility complex on T and NK cells, a receptor crucial for recognizing tumor-associated antigens and initiation of both adaptive (i.e., CTL) and innate (i.e., NK cells) immune responses [32]. Additionally, dormant cancer cells express low levels of immunogenic antigens, protecting them against T and NK cells [33]. Cancer cell-expressing immune-checkpoint molecules, such as programmed death ligand 1, bind to PD-1 on CTLs and thus suppress their antitumor activity [41]. Moreover, CTC-expressing CD47 binds to its receptor signal-regulatory protein-α on DCs and macrophages and thus inhibits their phagocytic function [42]. In the bloodstream, CTCs are decorated by platelets that protect them from antigen recognition by NK and T cells (Fig. 1.2), and platelet-secreted TGF-β- and IL-6/8 induce CTC proliferation and create immunosuppressive TME [32]. Finally, cancer cells cause the deletion of tumor-specific CTLs through apoptosis. For example, CTC-expressing Fas cell surface death ligand binds to FAS on T cells and initiates T cell apoptosis [43].

1.1.8 Reprogrammed metabolism

Altered metabolism is necessary to provide a selective advantage during tumors initiation and progression. Metabolic alterations emerge to (1) uptake the necessary nutrients, (2) adjust those nutrients with tumor-associated metabolic pathways, and (3) use metabolic reprogramming to affect the fate of cancer cells and other cells present in the TME. Cancer cell survival and proliferation and ATP generation require increased uptake of glucose and glutamine used as building blocks for catabolism and oxidation of macromolecules to reduce the power of biosynthetic reactions [44]. Moreover, glutamine with nitrogen-donation capacity contributes to the (de novo) biosynthesis of nucleotides and nonessential amino acids. Growth factor signaling and extracellular stimuli regulate glucose uptake; for example, ECM detachment leads to metabolic stress characterized by reduced glucose uptake, which negatively affects cell size and mitochondrial potential and decreases ATP generation [45,46]. Cancer cells encompass oncogenic alterations that result in hyperactivated PI3K/Akt/Ras signaling that promotes glucose influx through the expression of glucose transporter 1 and the first enzyme of the glycolytic pathway hexokinase to prevent glucose efflux into the extracellular space. Loss of c-Myc and Rb tumor suppressors associated with upregulation of the uptake and utilization of glutamine is critical for DNA replication in proliferative cancer cells [44].

Rapid tumor growth causes insufficient metabolic resources, with acquired mutations (e.g., mutant Ras, c-Src, and KRAS) providing alternate means for the uptake and lysosomal degradation of low-molecular-weight extracellular proteins (i.e., macropinocytosis) [47], engulfment and digestion of whole living cells (i.e., entosis), and capture of apoptotic bodies and extracellular macromolecules, all of which recover amino acids and allow for cancer cell survival and proliferation [44,48]. In another way, metastatic cancer cell-expressing fatty acid-binding protein 4 allows them to capture fatty acids directly released from neighboring normal adipocyte cells [49]. Upon nutrient starvation, cancer cells activate adaptive pathways such as autophagy or macroautophagy that provide nutrients and metabolites to ensure their surveillance [18]. Immune cell-expressing TNF-related apoptosis-inducing ligand (TRAIL) selectively bind to TRAIL-receptor or death receptor 4/5 (DR4/5) on cancer cells or CTCs and consequently activate downstream processes to induce apoptosis [50]. However, hypoxic stress activates autophagy, which induces DR4/5 endocytosis from CTC surfaces to protect them from apoptosis [51].

According to the Warburg effect, proliferative cancer cells utilize aerobic glycolysis rather than oxidative phosphorylation to fuel glucose, glutamine, and fatty acids for energy production and support growth by creating daughter cells [52]. However, quiescent tumor cells use those fuels for oxidation of carbon dioxide—i.e., oxidative phosphorylation (OXPHOS)—through the tricarboxylic acid cycle associated with reduced NADH and $FADH_2$ as well as ATP production utilized in basic cellular processes [44]. Hypoxia-driven HIF1α and c-Myc oncogene increase the expression of pyruvate dehydrogenase kinase 1, lactate dehydrogenase A, and monocarboxylate transporter, which conduct pyruvate to the Krebs cycle, convert pyruvate to lactate, and efflux lactate into the extracellular space, respectively [44,53]. Thus, metabolic plasticity allows tumor cells to adjust their metabolic phenotypes with TME components usually regulated at the gene expression level. Finally, the metabolic state of cancer cells influences the fate of other cells in the TME. For example, overproduced extracellular lactate attenuates immune-supportive T cells and DCs [54,55] but also promotes the emergence of M2 macrophages and even induces VEGF production and angiogenesis, which contribute to tumor invasiveness [56,57]. Thus, metabolic reprogramming is a contributor at various stages of tumorigenesis.

1.1.9 Natural acquisition of resistance

While cancer treatment is continuously evolving, intrinsic and acquired resistance to chemotherapy, radiotherapy, and hormone therapy leads to cancer, which is still the third-leading cause of death worldwide [58]. Naturally resistant cancer cells are attributed to preexisting mutant clones and switching of stemness markers that can drive tumor maintenance, metastasis, and recurrence. These properties first refer to the rare but heterogeneous populations of cancer cells in leukemia, termed tumor-initiating cells or cancer stem cells (CSCs) [59]. CSCs can originate from reprogramming of stem cells or somatic cells containing a specific load of genetic or epigenetic alterations, "drivers' mutation" (Fig. 1.3). Tumors in continuous evolution acquire "passenger mutations" that contribute to tumor growth, resistance, and survival due to the emergence of different types of CSCs [60]. CSCs have self-renewal (producer of CSCs and cancer cells) and differentiation (convert into the varied cell types in a TME) abilities governed by stem cell signaling pathways including Wnt/β-catenin, Notch, and Hedgehog (Hh) [18]. These pathways, aside from stemness properties and CSC markers such as NANOG, aldehyde dehydrogenase 1 (ALDH1), OCT4, SOX2, CD44, CD133, and CD55, provide cancer cells with resistance to current anticancer therapies [60].

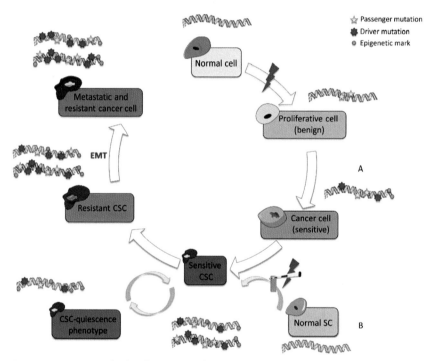

Figure 1.3 Origin and role of CSCs in malignancy. CSCs can arise from mutation-driven differentiative reprogramming of (A) cancer cells or (B) adult stem cells. CSCs provide the survival advantages of tumor growth, metastasis, and cancer resistance. *CSC*, cancer stem cell; *SC*, stem cell. *(Copy from Garcia-Mayea Y, Mir C, Masson F, Paciucci R, Lleonart ME. Insights into new mechanisms and models of cancer stem cell multidrug resistance. Semin Cancer Biol 2020;60:166—180).*

Therapy-resistant somatic cancer cells and CSCs share many resistance mechanisms traits. In CSCs, these mechanisms are implicated by various intrinsic and extrinsic factors. Briefly, intrinsic factors include ATP-binding cassette (ABC) transporters, stemness capacity, dormancy, protective autophagy, enhanced DNA repair activity, EMT, metabolic plasticity, signaling pathways, apoptosis and immune evasion, and drug detoxification enzymes.

Extrinsic factors involve diverse effectors in TME such as hypoxia, metabolic reprogramming, stomal and immune cells presented in CSC niches and the ECM [18,60,61]. ABC transporters are charged by ATP to efflux chemotherapeutic agents and reduce drug accumulation within the cells and thus induce multidrug resistance phenotype in CSCs [61]. Hypoxia and then HIFs conduct CSCs into dormancy state by activating cell cycle checkpoints, upregulation of ABC transporters, antiapoptotic factors, and stemness factors, hence protecting CSCs against anti-proliferative chemo agents [62,63]. ALDH acts as a drug detoxification enzyme and catalyzes the oxidation of cyclophosphamide or paclitaxel alkylating agents into weak components and thus protects CSCs from drug-induced DNA damage and apoptosis [18]. Interaction between ECM components, hyaluronic acid, and CD44v6 on CSCs implicates stemness that increases CSC self-renewal, metastasis, and cisplatin resistance in human cancers [64]. Considering TME cells, TAM-derived TGF-β and IL-17 induce Treg activation for cerate immunosuppressive TME and ignite EMT and self-renewal features of CSCs [65,66]. CSCs resist agent-inducer DNA damage involving cell cycle arrest, promotion of DNA damage response, and activation of the efficient ROS-scavenging system [18,61]. CSCs can overcome death signals through (1) upregulation of cell cycle checkpoint proteins, antiapoptotic proteins, and inhibitors of apoptosis, and (2) downregulation of proapoptotic proteins or death receptors [18,60,61]. Finally, CSCs meet their metabolic demands through (1) inducing angiogenesis to provide oxygen and nutrients, (2) shifting phenotypes that display metabolic differences, and (3) using alternative energy supplies [61]. For example, treatment with metformin as an OXPHOS inhibitor leads to shifting of OXPHOS to glycolysis in CSCs and thus protects them from such therapy [67]. Overall, the pathways mentioned here provide an integrated picture of the inherent and acquired resistant behaviors of CSCs.

References

[1] Paul D. The systemic hallmarks of cancer. J Cancer Metastasis Treat 2020;6.

[2] Nowell PC. The clonal evolution of tumor cell populations. Science 1976;194(4260):23—8.

[3] Hanahan D, Weinberg RA. The hallmarks of cancer. Cell 2000;100(1):57—70.

[4] Hanahan D, Weinberg RA. Hallmarks of cancer: the next generation. Cell 2011;144(5):646—74.

[5] Fouad YA, Aanei C. Revisiting the hallmarks of cancer. Am J Cancer Res 2017;7(5):1016.

[6] Alkhazraji A, Elgamal M, Ang SH, Shivarov V. All cancer hallmarks lead to diversity. Int J Clin Exp Med 2019;12(1):132—57.

[7] Hong W, Wu Q, Zhang J, Zhou Y. Prognostic value of EGFR 19-del and 21-L858R mutations in patients with non-small cell lung cancer. Oncol Lett 2019;18(4):3887—95.

[8] Holderfield M, Deuker MM, McCormick F, McMahon M. Targeting RAF kinases for cancer therapy: BRAF-mutated melanoma and beyond. Nat Rev Cancer 2014;14(7):455—67.

[9] Maertens O, Cichowski K. An expanding role for RAS GTPase activating proteins (RAS GAPs) in cancer. Adv Biol Regul 2014;55:1—14.

[10] Jiang BH, Liu LZ. PI3K/PTEN signaling in angiogenesis and tumorigenesis. Adv Cancer Res 2009;102:19—65.

[11] Coschi CH, Ishak CA, Gallo D, Marshall A, Talluri S, Wang J, et al. Haploinsufficiency of an RB-E2F1-Condensin II complex leads to aberrant replication and aneuploidy. Cancer Discov 2014;4(7):840—53.

[12] Soussi T, Wiman K. TP53: an oncogene in disguise. Cell Death Differ 2015;22(8):1239—49.

[13] Curto M, Cole BK, Lallemand D, Liu C-H, McClatchey AI. Contact-dependent inhibition of EGFR signaling by Nf2/Merlin. J Cell Biol 2007;177(5):893—903.

[14] Okada T, Lopez-Lago M, Giancotti FG. Merlin/NF-2 mediates contact inhibition of growth by suppressing recruitment of Rac to the plasma membrane. J Cell Biol 2005;171(2):361—71.

[15] Shaw RJ. Tumor suppression by LKB1: SIK-ness prevents metastasis. Sci Signal 2009;2(86):pe55—.

[16] Liu Z-G, Jiao D. Necroptosis, tumor necrosis and tumorigenesis. Cell Stress 2020;4(1):1.

[17] Galluzzi L, Pietrocola F, Bravo-San Pedro JM, Amaravadi RK, Baehrecke EH, Cecconi F, et al. Autophagy in malignant transformation and cancer progression. EMBO J 2015;34(7):856—80.

[18] Dianat-Moghadam H, Heidarifard M, Jahanban-Esfahlan R, Panahi Y, Hamishehkar H, Pouremamali F, et al. Cancer stem cells-emanated therapy resistance: implications for liposomal drug delivery systems. J Control Release 2018;288:62—83.

[19] Scaffidi P, Misteli T, Bianchi ME. Release of chromatin protein HMGB1 by necrotic cells triggers inflammation. Nature 2002;418(6894):191—5.

[20] Grivennikov SI, Greten FR, Karin M. Immunity, inflammation, and cancer. Cell 2010;140(6):883—99.

[21] Wang H-H, Wu Z-Q, Qian D, Zaorsky NG, Qiu M-H, Cheng J-J, et al. Ablative hypofractionated radiation therapy enhances non-small cell lung cancer cell killing via preferential stimulation of necroptosis in vitro and in vivo. Int J Radiat Oncol Biol Phys 2018;101(1):49—62.

[22] Hassannia B, Vandenabeele P, Berghe TV. Targeting ferroptosis to iron out cancer. Cancer Cell 2019;35(6):830—49.

[23] Guterres AN, Villanueva J. Targeting telomerase for cancer therapy. Oncogene 2020;39(36):5811—24.

[24] Geserick C, Tejera A, Gonzalez-Suarez E, Klatt P, Blasco M. Expression of mTert in primary murine cells links the growth-promoting effects of telomerase to transforming growth factor-β signaling. Oncogene 2006;25(31):4310—9.

[25] Ghosh A, Saginc G, Leow SC, Khattar E, Shin EM, Yan TD, et al. Telomerase directly regulates NF-κB-dependent transcription. Nat Cell Biol 2012;14(12):1270—81.

[26] Liu Z, Li Q, Li K, Chen L, Li W, Hou M, et al. Telomerase reverse transcriptase promotes epithelial—mesenchymal transition and stem cell-like traits in cancer cells. Oncogene 2013;32(36):4203—13.

[27] Lü M-H, Liao Z-L, Zhao X-Y, Fan Y-H, Lin X-L, Fang D-C, et al. hTERT-based therapy: a universal anticancer approach. Oncol Rep 2012;28(6):1945—52.

[28] Li T, Kang G, Wang T, Huang H. Tumor angiogenesis and anti-angiogenic gene therapy for cancer. Oncol Lett 2018;16(1):687—702.

[29] Hanahan D, Folkman J. Patterns and emerging mechanisms of the angiogenic switch during tumorigenesis. Cell 1996;86(3):353—64.

[30] Lugano R, Ramachandran M, Dimberg A. Tumor angiogenesis: causes, consequences, challenges and opportunities. Cell Mol Life Sci 2020;77(9):1745—70.

[31] Dianat-Moghadam H, Mahari A, Heidarifard M, Parnianfard N, Pourmousavi-Kh L, Rahbarghazi R, et al. NK cells-directed therapies target circulating tumor cells and metastasis. Cancer Lett 2021;497:41—53.

[32] Dianat-Moghadam H, Azizi M, Eslami-S Z, Cortés-Hernández LE, Heidarifard M, Nouri M, et al. The role of circulating tumor cells in the metastatic cascade: biology, technical challenges, and clinical relevance. Cancers 2020;12(4):867.

[33] Fares J, Fares MY, Khachfe HH, Salhab HA, Fares Y. Molecular principles of metastasis: a hallmark of cancer revisited. Signal Transduct Target Ther 2020;5(1):28.

[34] Kim M-Y, Oskarsson T, Acharyya S, Nguyen DX, Zhang XH-F, Norton L, et al. Tumor self-seeding by circulating cancer cells. Cell 2009;139(7):1315—26.

[35] Liu T, Ma Q, Zhang Y, Wang X, Xu K, Yan K, et al. Self-seeding circulating tumor cells promote the proliferation and metastasis of human osteosarcoma by upregulating interleukin-8. Cell Death Dis 2019;10(8):1—13.

[36] Gkountela S, Castro-Giner F, Szczerba BM, Vetter M, Landin J, Scherrer R, et al. Circulating tumor cell clustering shapes DNA methylation to enable metastasis seeding. Cell 2019;176(1—2):98—112. e14.

[37] Zamarron BF, Chen W. Dual roles of immune cells and their factors in cancer development and progression. Int J Biol Sci 2011;7(5):651.

[38] Vinay DS, Ryan EP, Pawelec G, Talib WH, Stagg J, Elkord E, et al. Immune evasion in cancer: mechanistic basis and therapeutic strategies. Semin Cancer Biol 2015;35:S185—98.

[39] Maeda H, Shiraishi A. TGF-beta contributes to the shift toward Th2-type responses through direct and IL-10-mediated pathways in tumor-bearing mice. J Immunol 1996;156(1):73—8.

[40] Dunn GP, Old LJ, Schreiber RD. The three Es of cancer immunoediting. Annu Rev Immunol 2004;22:329—60.

[41] Ock C-Y, Kim S, Keam B, Kim M, Kim TM, Kim J-H, et al. PD-L1 expression is associated with epithelial-mesenchymal transition in head and neck squamous cell carcinoma. Oncotarget 2016;7(13):15901.

[42] Steinert G, Schölch S, Niemietz T, Iwata N, García SA, Behrens B, et al. Immune escape and survival mechanisms in circulating tumor cells of colorectal cancer. Cancer Res 2014;74(6):1694—704.

[43] Gruber I, El Yousfi S, Dürr-Störzer S, Wallwiener D, Solomayer E, Fehm T. Down-regulation of CD28, TCR-zeta (ζ) and up-regulation of FAS in peripheral cytotoxic T-cells of primary breast cancer patients. Anticancer Res 2008;28(2A):779—84.

[44] Pavlova Natalya N, Thompson Craig B. The emerging hallmarks of cancer metabolism. Cell Metabol 2016;23(1):27—47.

[45] Rethinking the regulation of cellular metabolism. In: Thompson C, editor. Cold Spring Harbor symposia on quantitative biology. Cold Spring Harbor Laboratory Press; 2011.

[46] Extracellular matrix regulation of metabolism and implications for tumorigenesis. In: Grassian A, Coloff J, Brugge J, editors. Cold Spring Harbor symposia on quantitative biology. Cold Spring Harbor Laboratory Press; 2011.

[47] Commisso C, Davidson SM, Soydaner-Azeloglu RG, Parker SJ, Kamphorst JJ, Hackett S, et al. Macropinocytosis of protein is an amino acid supply route in Ras-transformed cells. Nature 2013;497(7451):633—7.

[48] Palm W, Park Y, Wright K, Pavlova NN, Tuveson DA, Thompson CB. The utilization of extracellular proteins as nutrients is suppressed by mTORC1. Cell 2015;162(2):259—70.

[49] Nieman KM, Kenny HA, Penicka CV, Ladanyi A, Buell-Gutbrod R, Zillhardt MR, et al. Adipocytes promote ovarian cancer metastasis and provide energy for rapid tumor growth. Nat Med 2011;17(11):1498.

[50] Dianat-Moghadam H, Heidarifard M, Mahari A, Shahgolzari M, Keshavarz M, Nouri M, et al. TRAIL in oncology: from recombinant TRAIL to nano- and self-targeted TRAIL-based therapies. Pharmacol Res 2020;155:104716.

[51] Twomey JD, Zhang B. Circulating tumor cells develop resistance to TRAIL-induced apoptosis through autophagic removal of death receptor 5: evidence from an in vitro model. Cancers 2019;11(1):94.

[52] Warburg O, Wind F, Negelein E. The metabolism of tumors in the body. J Gen Physiol 1927;8(6):519.

[53] Wahlström T, Henriksson MA. Impact of MYC in regulation of tumor cell metabolism. Biochimica et Biophysica Acta (BBA)-Gene Regul Mech 2015;1849(5):563—9.

[54] Fischer K, Hoffmann P, Voelkl S, Meidenbauer N, Ammer J, Edinger M, et al. Inhibitory effect of tumor cell—derived lactic acid on human T cells. Blood 2007;109(9):3812—9.

[55] Gottfried E, Kunz-Schughart LA, Ebner S, Mueller-Klieser W, Hoves S, Andreesen R, et al. Tumor-derived lactic acid modulates dendritic cell activation and antigen expression. Blood 2006;107(5):2013—21.

[56] Stern R, Shuster S, Neudecker BA, Formby B. Lactate stimulates fibroblast expression of hyaluronan and CD44: the Warburg effect revisited. Exp Cell Res 2002;276(1):24—31.

[57] Sonveaux P, Copetti T, De Saedeleer CJ, Végran F, Verrax J, Kennedy KM, et al. Targeting the lactate transporter MCT1 in endothelial cells inhibits lactate-induced HIF-1 activation and tumor angiogenesis. PloS One 2012;7(3):e33418.

[58] Bray F, Ferlay J, Soerjomataram I, Siegel RL, Torre LA, Jemal A. Global cancer statistics 2018: GLOBOCAN estimates of incidence and mortality worldwide for 36 cancers in 185 countries. CA — Cancer J Clin 2018;68(6):394—424.

[59] Lapidot T, Sirard C, Vormoor J, Murdoch B, Hoang T, Caceres-Cortes J, et al. A cell initiating human acute myeloid leukaemia after transplantation into SCID mice. Nature 1994;367(6464):645—8.

[60] Garcia-Mayea Y, Mir C, Masson F, Paciucci R, Lleonart ME. Insights into new mechanisms and models of cancer stem cell multidrug resistance. Semin Cancer Biol 2020;60:166—80.

[61] Najafi M, Mortezaee K, Majidpoor J. Cancer stem cell (CSC) resistance drivers. Life Sci 2019;234:116781.

[62] Schöning JP, Monteiro M, Gu W. Drug resistance and cancer stem cells: the shared but distinct roles of hypoxia-inducible factors HIF 1α and HIF 2α. Clin Exp Pharmacol Physiol 2017;44(2):153−61.

[63] Liao W-L, Lin S-C, Sun HS, Tsai S-J. Hypoxia-induced tumor malignancy and drug resistance: role of microRNAs. Biomark Genomic Med 2014;6(1):1−11.

[64] Bourguignon LY. Matrix hyaluronan promotes specific MicroRNA upregulation leading to drug resistance and tumor progression. Int J Mol Sci 2016;17(4):517.

[65] Xiang T, Long H, He L, Han X, Lin K, Liang Z, et al. Interleukin-17 produced by tumor microenvironment promotes self-renewal of CD133+ cancer stem-like cells in ovarian cancer. Oncogene 2015;34(2):165−76.

[66] Chanmee T, Ontong P, Konno K, Itano N. Tumor-associated macrophages as major players in the tumor microenvironment. Cancers 2014;6(3):1670−90.

[67] Banerjee A, Birts CN, Darley M, Parker R, Mirnezami AH, West J, et al. Stem cell-like breast cancer cells with acquired resistance to metformin are sensitive to inhibitors of NADH-dependent CtBP dimerization. Carcinogenesis 2019;40(7):871−82.

CHAPTER 2

Nanotechnology for cancer theranostics

1. Nanoscience and nanotechnology

Research advances continue to result in the development of new fields. Some come, and some go. Others develop into well-recognized entities with promising applications for medicine and public health. Among the most dynamic emerging fields is nanotechnology, including nanomedicine. Nanotechnology is expected to propagate and further revolutionize medicine and may well have applications that span all aspects of disease prevention, diagnosis, and treatment. This growing realization has attracted investment in the field. By 2008, the United States had invested an estimated $3.7 billion in nanotechnology. Outside the United States, the European Commission authorized $1.7 billion in its Sixth Framework Programme for Research and Technological Development during the same period. In China, nanotechnology research has taken center stage, spurred by international conferences that have been held since 1990. These examples and others show that nanotechnology is growing. Nanotechnology offers the potential for a broad range of applications with impacts on the practice of medicine and provision of public health services [1,2]. Not all advances in nanotechnology occur in the medical and health sciences. Nanotechnology principles and concepts have been developed in other rich and highly relevant fields. Indeed, scientists with diverse backgrounds (e.g., chemists, physicists, toxicologists, material scientists, and engineers) have converged to work on understanding the elements of nanotechnology.

1.1 Nanotechnology and nanomedicine

Nanotechnology's origins are often assigned to a concept advanced by Nobel laureate Richard P. Feynman. In a 1959 lecture at the California Institute of Technology, he stated the following:

Targeted Cancer Imaging
ISBN 978-0-12-824513-2
https://doi.org/10.1016/B978-0-12-824513-2.00003-6

There is plenty of room at the bottom. . . . Many of the cells are very tiny, but they are very active: they manufacture various substances; they walk around; they wiggle: and they do all kinds of marvelous things—all on a very small scale. Also, they store information. Consider the possibility that we too can make a thing very small which does what we want—that we can manufacture an object that maneuvers at that level! [3].

Nanoparticles (NPs) are nothing new. They occur naturally and are also formed during ignition processes. Living systems are inherently composed of nanoscale building blocks, and most investigations into molecular biology are already pursued in nanoscale. What is new concerning nano-materials is the ability to control production and make engineered NPs. For many decades, chemists have been learning to control the arrangement of low numbers of atoms inside molecules, resulting in the ability to create high-performance plastics, effective drugs, and other purpose-designed materials. Nowadays, nanotechnology encompasses the study, manipula-tion, and application of material properties at atomic scales. "Nano" derives from the Greek noun nanos, meaning dwarf. A nanometer (nm) is 1×10^{-9} (one billionth of a meter), which is the length of 10 hydrogen atoms placed side by side.

Nanotechnology relates to the organization of atoms and molecules within a size range of 1—100 nm. The width of a DNA molecule is 2.5 nm. At the nanoscale, materials do not behave like gases, liquids, or solids but rather exist within the world of quantum physics (the branch of physics that uses quantum theory to describe and predict the properties of physical systems). Nanomedicine refers to applications of nanoscience to diagnose, monitor, treat, and control biological systems. Nanomedicine includes nanostructures that act as biological mimics, nanofibers, and polymeric nanostructures, such as biomaterials, sensors, and diagnostic systems [4,5]. Nanoscience and nanotechnology require a particularly wide spectrum of skills and knowledge. For this, many interdisciplinary research programs and other efforts have been established to promote collaboration among rele-vant fields (e.g., the NCI's Nanobiology Think Tank, an outreach program for promoting collaboration and disseminating knowledge to enhance and develop novel creative ideas for biomedical applications). In nanomedicine, prime fields of investigation include nano-electronics, which aims to bring molecular manufacturing to fruition. Nanobiotechnology is another field of investigation.

The thrust here is to combine nanoscale engineering with biology to manipulate living systems directly or build biologically inspired nanomaterials and devices at the molecular level. Efforts to combine biology and nanotechnology have focused on fusing useful biomolecules and chemically synthesized NPs in arrangements that do everything from emitting light to storing tiny bits of magnetic data. The output of this and related nanotechnology research has emerged as a benchtop tool for the rapid diagnosis of infectious diseases and detection of tiny genetic mutations. Other nanomedicine projects aim to deliver and target diagnostic, pharmaceutical, and therapeutic agents by intravenous and interstitial routes of administration with nanostructures [6]. Similarly, biosensing has attracted the attention of nanomedicine researchers; it requires the interfacing of a nanodevice with a patient in some way [7]. One approach is nanotubes inserted through the skin to continuously monitor blood glucose levels [8]. These fabricated devices have important clinical uses. In orthopedics, nanotechnology will lead to new diagnosis procedures, better prophylactic techniques, and new nanomaterials for bone repair [9]. In oncology, nanotechnology provides oncologists with new tools for tracking and targeting cell surface receptors. Silicon-based nanowire arrays have been used to detect overexpressed cancer biomarkers in blood [10]. In addition, nanotechnology provides potential strategies to overcome anticancer drug resistance. Studies have shown that nanocarriers can deliver anticancer drugs into cells without triggering the p-glycoprotein pump. Another approach targets tumor cells with gold nanostructures, which with NIR irradiation, trigger heat that kills tumor cells [11]. Another example of applying nanotechnology is the search for an accurate and noninvasive method for detecting lymph node metastases in cancer. These investigations indicate that magnetic NPs with high-resolution MRI allow detection of small and otherwise undetectable lymph node metastases in patients [12]. In lymph nodes, nanostructures are internalized by macrophages, and these intracellular NPs cause changes in magnetic properties [13]. Added to this list of developments, in 2006 the surgery branch of NCI's Center for Cancer Research announced the beginning of clinical trials with a novel nanoformulation of tumor necrosis factor protein. There are parallel concerns of shortening clinical trials and the regulatory pathway for new drugs. This interest has turned to nanotechnology to develop nanodevice reporters that could provide real-time data and enable clinicians to determine within hours or days of treating a patient whether a candidate drug was killing cancer cells as intended [14]. In addition, the NIH Roadmap for Medical Research was designed to enhance the nation's clinical research enterprises and includes significant nanomedicine initiatives. These initiatives cause nanotechnology that improves biomedical strategies. This includes developing nanoscale laboratories for diagnostic and drug discovery. Another interesting approach is using nanotechnology in system biology to understand how systems communicate with one another and how cells communicate among/within themselves and with all other cell systems within living systems [15]. In environmental health, nanotechnology can provide tools to determine external and internal exposures in real time, assess risk, and link exposures to disease. In addition, nanomaterials designed to provide nontoxic, nonflammable, "reduced risk" products are currently on the market. Examples include self-cleaning or self-sterilizing surfaces with important applications in healthcare facilities. In summary, the promising application of nanotechnology may well contribute important tools for promoting health, preventing disease, and improving quality of life for all segments of the population [1,5].

1.2 Nanomedicine in cancer bioimaging

Early-state cancer detection is associated with patient prognosis, as it allows immediate therapeutic interventions to prevent cancer progression. Nevertheless, this does not always occur, as cancers are usually detected after they have already formed and reached a detectable, adequately large, and palpable size. In such cases, to confirm that the cells are cancerous or precancerous, bioimaging and biopsy (to biomarker screen) may confirm detection. Heterogeneity within the cancer lesion is another challenge that complicates cancer detection. Thus, at the cancer site, cancerous cells exist beside precancerous cells and noncancerous cells (e.g., stromal and immune cells). This heterogeneity hinders the accurate evaluation of specimens at the molecular and cellular levels. Nanomedicine promises to improve cancer detection by enhancing imaging and screening techniques through the use of nanofabrication and nanosystems that create new early detection platforms. In addition to these improvements, nanomedicine creates conditions for researchers to design the targetable platform and multimodal imaging and combine detection and therapy agents within one system (theranostic) [16–18].

Quantitative measuring of imaging contrast agents is desired to establish automated algorithms and guidelines for detecting, real-time monitoring, and evaluating cancer therapy outcomes. Imaging modalities based on contrast agents fall within four groups: (1) optical, such as near-infrared fluorescent imaging, resonance energy transfer with Raman, optical coherence tomography, and photoacoustic imaging, (2) magnetic, such as magnetic resonance imaging (MRI), magnetomotive imaging, and magnetic particle imaging, (3) acoustic ultrasound imaging, and (4) nuclear, such as computed tomography (CT), positron emission tomography, single-photon emission-computed tomography, and Υ-imaging [19]. These current clinical modalities fall under the category of tomographic imaging, which relies on deep-penetrating radiation to probe structural and functional information of the imaged area. The clinical application of different imaging modalities has been investigated (see review Refs. [19,20]). Despite all the benefits, these approaches face limitations for clinical utility. For example, optical imaging presents advantages such as being inexpensive, having high spatial resolution, and avoiding the harm associated with ionizing radiation. Yet the limited penetration depth of its related contrast agents restricts its clinical application to imaging-guided surgery, colonoscopy, and endoscopy [21,22]. The advantages and disadvantages of these imaging strategies are summarized in Table 2.1.

Regarding these limitations, nanotechnology is a promising field at the forefront of cancer detection research. Considerable efforts have been made to create various targeted molecular imaging nanoplatforms with unique features and capabilities. Unlike conventional imaging techniques, tumor-selective imaging probes would deliver an optimized imaging agent to a specific target with high affinity, specificity, and sensitivity. Secondly, a lower but effective dosage of tumor-selective therapeutic nanoplatforms could efficiently localize within tumors with minimized systemic toxicity. Moreover, it is possible to monitor and confirm whether nanoplatform-based tumor-selective imaging probes have been properly delivered to the targeted site after injection. Compared with traditional imaging agents, the number of injected tumor-selective imaging probes that actually reach tumor sites can be quantitatively analyzed. In addition, biodistribution of these probes within the body can be monitored over a long period. So targeted nanoplatforms have good potential to significantly increase imaging contrast, enable cancer detection at earlier stages, and provide monitoring of responses to conventional tumor therapy or molecular targeted therapy [23].

Other benefits of using targeted nanoprobes are, firstly, that imaging nanoplatforms can be used to monitor the changes in the molecular microenvironment associated with tumors. Secondly, the integration of imaging and therapeutic capabilities provides a combined diagnostic therapy, termed a theranostic approach [24]. Theranostic systems can reduce

Table 2.1 Conventional imaging modalities used in the clinic.

Modality	Advantage	Limitation	Input signal type	Resolution (mm)	Penetration depth
CT	Rapid, accurate, low cost, reproducibility, widely available	Limited resolution, imaging interpretation difficult, ionizing radiation	X-ray	25–200 μm (preclinical), 0.5 –1 mm (clinical)	Unlimited
MRI	Soft-tissue contrast, high resolution, customizable molecular targeting, cell tracking	High cost, large equipment requirement, limited sensitivity, requires contrast agent	Radio frequency	25–100 μm (preclinical), ∼1 mm (Clinical)	Unlimited
USI	Rapid, accurate, low cost, reproducibility, widely available	Limited resolution, image interpretation difficult, artifacts common	Sound waves	10–100 μm (preclinical), 1 –2 cm (clinical)	10 ms
PET	Quantification of metabolism and blood flow, high sensitivity, many radionuclide tracers available	High cost, limited availability, large equipment requirement, short tracer half-life, single process evaluation	Radionuclide (positrons)	<1 mm (preclinical), ∼5 mm (clinical)	Unlimited
PAT	Reduce tissue scattering, high resolution, nonionizing/ nonradioactivity, no acoustic noise, high penetration depth, and high resolution	Limited path length, dependence on temperature, weak absorption at short wavelengths	Light	5 μm–1 mm (depth-dependent)	<6 cm
SPECT	3-D imaging, widely available, highly sensitive, simultaneous imaging of multiple processes	Limited temporal resolution, few radionuclide tracers	Radionuclide (γ-rays detected)	0.5–2 mm (preclinical), 8–10 mm (clinical)	Unlimited
NIFI	Low cost, widely available	Photobleaching, low quantum yield, shallow tissue penetration	Ultraviolet to near-infrared light	2–3 mm	<2 cm

CT, computed tomography; *MRI*, magnetic resonance imaging; *NIFI*, near-infrared imaging; *PAT*, photoacoustic imaging; *PET*, positron emission tomography; *SPECT*, single–photon emission–computed tomography; *USI*, ultrasound imaging.

toxicity, enhance selectivity and targeting, generate data for diagnostics, and enhance therapeutic efficiency [20].

1.2.1 Engineering targeted nanostructures for cancer imaging

To maximize the diagnostic precision of a nanosystem imaging agent, it must be designed with a method that maximizes its transport to the cancer site. Fortunately, design flexibility is the substantial advantage of a targeted nanocarrier. A wealth of materials, including metals, polymers, and lipids, enable the production of NPs for many clinical imaging modalities (e.g., CT, MRI). Top-down fabrication methods, such as particle replication in the nonwetting template method and the dip-pen nanolithography method, produce nanosized particles with a wide variety of shapes, structures, and sizes from a variety of different bulky materials [25,26]. Furthermore, plenty of activable chemistries groups enable the conjugation of ligands and polymers to nanocarriers, which improve targetability, biocompatibility, and specificity. Targeted therapeutic nanoplatform must undergo rigorous preclinical testing (including safety and toxicity studies) and validation to be approved by the FDA. In comparison with therapeutics nanosystems, consideration of safety is even more important with imaging nanosystems because they could be administered to healthy persons. With the synthesis flexibility of targeted imaging nanosystems of different sizes, shapes, and materials, nanosystem design can be optimized to maximize the specificity, selectivity, and sensitivity of cancer targeting. By fine-tuning a nanosystem's shape, surface chemistry, and size, its targeting abilities and margination can be significantly improved. Margination is defined as the ability of a nanosystem to escape the blood flow and move to the vessel wall, which is a required process to target a nanosystem to cancer. If a nanosystem radial movement in a vessel is limited, the opportunity for binding interactions between the nanosystem and vessel wall will also be restricted. Modification of a nanosystem's surface chemistry, size, and shape also affect its binding avidity. An ideal targeted imaging nanoplatform for tumor imaging would be produced with the following specifications: (1) high rate of margination, (2) strong binding avidity to tumor tissue, and (3) rapid internalization by the targeted tumor cells. Creating a targeted imaging nanoplatform with all three of these characteristics is challenging because modification of one parameter may enhance one or more other characteristics. In this context, targeted imaging nanoplatforms must be designed with consideration of all these biochemical and biophysical phenomena. Targeted imaging nanoplatform performance is traditionally evaluated through independent in vitro internalization, in vivo pharmacokinetics, and biodistribution. A caveat in the majority of in vitro internalization studies is that they often neglect the effect of blood flow on nanoplatform transport to the vessel wall. Accurate assessment of in vitro nanoplatform targeting can be done in a parallel plate chamber and microfluidic setup that replicates vascular morphology and allows for control of local flow [27–29]. Similarly, nanoplatform in vivo studies primarily focus on the efficacy of elements (e.g., polymer) that extend circulation half-life and target ligands that improve tumor specificity without considering the uniqueness of different tumor tissue microenvironments. These studies not only neglect the importance of nanoplatform geometry but also fail to account for variability in fluid dynamics, which has been shown to be an important obstacle to nanoplatform delivery to the tumor site [30].

1.3 Nanomedicine in cancer vaccination

Nanomedicine offers many aids that can be leveraged to increase vaccine potency. One advantage of nanoplatform-based formulations is the ability to codeliver antigenic agents with an immunostimulatory adjuvant. This is very important for proper immune stimulation because colocalization of the two ingredients ensures quick generation of an immune response to the antigen [31]. To achieve codelivery, the adjuvant can be encapsulated into the nanoplatform core or functionalized onto the nanoplatform surface, or the nanoplatform itself can serve as a stimulus. Studies have shown that nanoplatform-based vaccines can significantly enhance the presentation and activation of CD^{8+} T cells by at least 500-fold compared with vaccination by soluble antigen alone. Animal groups treated with the nanovaccine completely rejected tumor growth and remained tumor-free for 40 days. Results of this study showed that when using a mixture of ovalbumin (OVA) and alum (a traditional adjuvant), there was little improvement in survival and no observable decrease in tumor growth [32]. Other examples of stimulatory nanostructures include small-sized proteoliposomes and calcium phosphate (as a mineral-based adjuvant) [33,34]. Adjuvant elements can be directly incorporated within a nanoplatforms matrix. For example, interbilayer cross-linked multilamellar vesicles (ICMVs) were formulated by cross-linking the head groups of lipid bilayers as adjacent [35]. The complex structure of ICMVs can incorporate a high number of OVA proteins as well as monophosphoryl lipid A, a lipid-like immunostimulatory adjuvant.

Immunization with ICMVs evokes a strong humoral response with a substantial increase in antibody levels (about 1000-fold compared with immunizations using soluble OVA antigen). ICMVs can generate strong cell-mediated immunity by eliciting a potent CD^{8+} T cell response against OVA (the magnitude was 14 times stronger than when using soluble antigen). ICMV-based nanovaccines have also been utilized in conjunction with other adjuvants such as polyIC (a TLR3 agonist), which is used to produce antigen-specific effector memory T cells at the mucosal surface (many pathogens infiltrate their hosts through these barriers) [36].

Another strategy for incorporating adjuvants into vaccination nanoplatforms directly functionalizes them onto nanoplatform surfaces. For example, synthetic high-density lipoprotein-mimicking nanodiscs (HDLMNs) have been used for immunotherapeutic applications [37]. In one study, HDLMN surfaces were functionalized with adjuvant CpG (a TLR9 agonist) and tumor antigens via reduction-sensitive linkages. This HDLMN-based vaccination nanoplatform, when injected into mice, could produce an increase (41-fold) in antigen-specific CD8+ T cells compared with antigen-linked HDLMNs mixed with free CpG (as control groups). When HDLMN-based vaccine was used in conjunction with checkpoint inhibitors to generate a response against a mutated neoantigen in a tumor model, complete regression was achieved in 87.5% of the mice. As a comparison, only 25% survival was achieved with soluble neoantigen and adjuvant formulation. The HDLMN-based vaccine has also been combined with chemotherapy drugs to achieve impressive therapeutic results against cancers [38].

Another advantage of nanoplatform-based vaccines is the enhanced bioavailability of the cargo. By conjugating or encapsulating antigen and adjuvant to a nanocarrier, the effective agent can be more effective and protected from host interactions during transport. Within the blood circulatory system, a wide range of enzymes can cause the degradation of bioactive and biomolecular agents [39]. In addition, nanoplatform delivery systems can prevent the systemic toxicity often associated with the administration of adjuvant in its free form. Accordingly, by shielding their cargo from the surrounding environment, nanoplatforms can concurrently protect the host from nonspecific interactions that can cause deliberate side effects [40]. The protection imparted by nanocarriers can effectively prolong in vivo residence, which increases the probability of successful delivery to antigen-presenting cells. Furthermore, targeting agents can be conjugated onto the vaccination nanocarrier surface to enhance delivery to desired cell subsets. Some ligands (e.g., mannose) have a natural affinity for macrophages and dendritic cells (DCs) [41]. Lastly, nanoplatform-based formulations can strongly prolong immune stimulation as well as the controlled and sustained release of encapsulated cargos. Careful selection of the components used in constructing a vaccine's nanoformulation can ensure the slow release of the encapsulated cargos over time in a controlled manner, which has been shown to prolong the elevation of antibody and results in the production of a high level of effector memory T cells [42,43]. The size range of nanoplatforms is another factor that can enable improved delivery of a vaccine-effective agent. The nanoscale ranges of nanocarriers allow for efficient lymphatic drainage into the lymphoid organs where antigen uptake and processing occur [44–46]. In one example application of this effect, iron oxide–zinc oxide nanostructures with a core–shell structure were used for delivery of carcinoembryonic antigen [47]. These NPs had an average size of 15 nm, enabling those to effectively travel into lymph nodes. Once at their destination to the lymph nodes, the NPs could be taken up by DCs for processing to extract specific immunity against the antigen. When this nanoformulated vaccine was administered to mice, there was a 10-fold increase in splenic CD8+ T cells secreting IFN-γ (proinflammatory cytokine commonly correlated with the activation of cell-mediated immunity). Although the nanoformulated vaccine could not completely eradicate the tumor, vaccination with iron oxide–zinc oxide nanostructures delayed tumor growth and extended mean survival from 10.5 to 19.5 days. An added advantage of these nanoformulations was their inherent ability to be used for MRI. It is important to note that effective lymphatic drainage of nanostructures depends greatly on size. Research has found that after intradermal injection, 100 nm nanostructures were only 10% as efficient in accumulating at the draining lymph nodes when compared with 25 nm nanostructures [46]. Nanocarriers can also be designed for cytosolic delivery, which has important implications for improving vaccine performance. Antigen delivery to the cytosolic compartment is a unique way of activating CD^{8+} T cells via the MHC-I pathway. In the normal conditions, internalized antigens must be shuttled from the endosome to the cytosol to be presented in the context of MHC-I [48]. Nevertheless, targeted delivery of antigenic nanomaterial directly into the cytosol can allow CD^{8+} T cell stimulation while bypassing endogenous presentation mechanisms. For cancer vaccines, effective cytosolic delivery can benefit from both antigen and adjuvant, leading to simultaneous enhancement of antigen and improvement of immune stimulation [49]. In one study, synthetic polymeric nanostructures were used to deliver protein antigen OVA to the cytosol while simultaneously triggering the stimulator of the interferon gene pathway [50]. The nanocarriers were designed to be pH-sensitive, enabling them to disrupt the endosomal membrane and release their cargos into the cytosol prior to being degraded. Compared with a soluble protein antigen ovalbumin control, the nanoformulation could increase the frequency of protein antigen ovalbumin-specific CD^{8+} T cells by 29-fold. Ultimately, when used under a therapeutic setting, simultaneously triggering the stimulator of interferon gene pathway agonistic nanoformulations were capable of significantly controlling tumor growths. Other strategies for cytosolic delivery include employing hydrophobic nanocarriers and cationic nanocarriers to facilitate direct uptake [51,52].

1.4 Nanomedicine in cancer therapy

Cancer is one of the important causes of morbidity and mortality worldwide. Despite extensive and advanced research on new approaches in cancer therapy, current treatments are still limited to surgery, chemotherapy, immunotherapy, and radiotherapy. Therapeutic failures are related to drug resistance and pharmacological toxicity issues in most instances. Nanotechnology has substantial application in all areas of science and has extended its role in the field of medicines and health care. Advanced efforts have been made to develop novel nanocarrier-based delivery and smart delivery systems for effective cancer therapy [53]. Nowadays, using nanocarriers leads to increased therapeutic efficacy and tumor site concentrations of the drug; extended blood circulation half-time, distribution volume, cellular uptake, and half-life are important factors for an improved therapeutic window following clinical success. The focus is to develop a nanocarrier system that can make modifications at the gene level [54,55]. The most common examples of these nanocarriers include micellar, liposomal, dendritic, polymeric, solid lipid, nanoshell, magnetic, and metallic nanoparticles [56–58].

1.4.1 Nanocarriers in cancer therapy

Polymeric NPs are some of the most commonly used nanocarriers in cancer therapy. They are used to encapsulate therapeutic cargo and release it at the specific tumor site. The type of polymer and the ratio of hydrophobicity to hydrophilicity within the polymer are important factors responsible for the rate of drug release. Polyethylene glycol (PEG) consists of a glycol moiety (imparting hydrophilicity to the carrier), and it is readily degraded by hydrolysis. In addition, polylactic acid (PLA), a hydrophobic polymer, has a long half-life, as it is less prone to enzymatic degradation [59–61]. Due to the enhanced effectiveness of prepared polymeric NPs, combinations of PLA and poly(glycolic acid) were used in the development of a poly(lactic-co-glycolic acid) (PLGA) polymer having both hydrophobic and hydrophilic properties. In this nanocarrier, altering the ratios of lactic acid and glycolic acid can modify drug release from the carriers and also alter the pattern and time of degradation [59]. In most cases, controlled release systems provide optimal drug release for a prolonged period, thus minimizing repeated dosing of patients [62]. However, while designing any drug delivery system, it is primarily taken into care to deliver drug within its therapeutic range starting from initial burst release to the end releasing, this causing reduced toxicity and turn increases safety, and finally causing the increasing patient compliance [59,62]. In the case of cancers, continuous maintenance doses are required to increase the effectiveness of treatment as the efficacy of therapy is directly proportional to the death of a number of cancerous cells without affecting normal cells. In some cases, an IV bolus can be administered to retard drug release from the nanocarrier system to the systemic circulation [62]. However, degradation of the polymeric nanocarrier determines the fate of drug release from the carrier system. It can occur by surface erosion, swelling of the polymer, or diffusion by bulk, or it can be a condition-dependent or time-dependent release system. Drug release may be constant or cyclic for a long period [59,62,63]. The surface of the polymeric nanocarrier can be functionalized to specific targeting of cancer cells or for prolonging the half-life of the delivery system by preventing its uptake by the reticuloendothelial system (RES). Binding of plasma protein is the initial step in the uptake of nanocarriers by the RES, which causes major loss of administered dose. These surface-modified PEGylated nanocarriers are often referred to as the "stealth effect," and the surface is mainly modified to improve pharmacokinetics by escaping surveillance of the RES. Several other examples of polymeric marketed nanoformulation include Nutropin AQ Depot, Vivitrol, Risperdal Consta, and Arestin [54,57,64]. In addition, active targeting increases specific cell targeting. For example, a transferrin-conjugated PTX-loaded PLGA nanocarrier showed cellular uptake of up to 35% compared with untargeted PTX PLGA NPs, which showed maximum uptake of only 15%. Similarly, transferrin-conjugated nanocarriers showed greater antiproliferative activity than untargeted nanocarriers [57,59,63].

Lipid-based nanocarriers are another suitable candidate for the design of drug delivery nanocarriers. Liposomes are NPs with a bilayer membrane made up of synthetic or natural lipids consisting of multilamellar and unilamellar vesicles. These structures can load both hydrophilic as well as hydrophobic drugs simultaneously within them; this specific feature is due to the presence of amphiphilic lipids involving the hydrophilic head and hydrophobic tail [65]. In this carrier, hydrophilic drugs can be encapsulated within the hollow core, and hydrophobic drugs can be encapsulated in the lipid bilayer. Some of the liposomal-based formulations have been clinically approved by regulatory agencies for routine use. Doxil was the first liposomal-based formulation of DOX and was approved by FDA [66]. The surface functionalization of Doxil with methoxy polyethylene glycol prevents RES uptake, thus prolonging the circulation time of the drug in systemic circulation. DaunoXome is another liposomal-based formulation and has been approved against Kaposi' sarcoma and other cancers [67]. A series of modifications can be made to develop a pH-responsive liposomal-based formulation that can release drugs from the nanocarrier system into the acidic environment [68–70].

Magnetic nanoparticles (MNPs) are another class of nanocarriers observed to have diagnostic and therapeutic applications. MNPs are being developed for diagnostic and therapeutic applications such as MRI contrast agents and MNP-based targeted drug delivery [71]. The surface of the MNP nanocarrier is conjugated with targeting ligand or imaging modalities. Therapeutic agents can be conjugated or chemically bonded to the carrier surface [72,73]. MNPs are stable, but they must be coated with small organic molecules such as dextran, PEG, and poly(ethyleneimine) and polymers like chitosan because MNPs agglomerate, thus reducing specific targeted cell interactions. MNPs are conjugated with different peptides, aptamers, antibodies, proteins, and small organic molecules to effectively target particular types of cancer cells [74,75]

Nanotechnology, besides improving routine cancer therapy methods, has opened a new therapy window to better treat cancer. Photothermal therapy (PTT), a noninvasive therapeutic strategy in which photon energy is converted into heat to destroy cancer cells, has been used to treat cancer in varying degrees over the past few decades [76]. Heating sources including microwaves and near-infrared, radiofrequency, and ultrasound waves are used to induce moderate increases in temperature at specific target sites to destroy cancerous cells (clinically termed hyperthermia). In the last decade, synthetic organic dye molecules, such as indocyanine green, and naphthalocyanine have been used to enhance photothermal effects. Because dye molecules photobleach quickly, PTT has not been widely used in the clinic. Nowadays, PTT has attracted new interest for cancer therapy because of the generation of a new class of photothermally sensitized agents. Gold at the nanoscale shows superior light absorption efficiency over conventional dye molecules. Upon irradiation to radiation, strong surface fields are induced from the coherent excitation of the electrons in the gold nanostructures. The rapid relaxation of these excited electrons produces strong localized heat capable of destroying targeted cancer cells via hyperthermia [77].

1.4.2 Combination therapy

Combining two or more therapeutic methods to specifically target cancer-inducing cells is a cornerstone of cancer therapy [78]. Although the monotherapy strategy is still a common treatment modality for many different cancers, this conventional method is generally deemed less effective than the combination therapy approach. Conventional mono-therapeutic methods nonselectively target actively proliferating cancer cells, which ultimately leads to the destruction of both healthy and cancer cells. Chemotherapy can be toxic, with multiple side effects and risks, and greatly reduce immune system effectiveness by its effects on bone marrow cells [79]. Although combination therapy can be toxic if one of the agents used is chemotherapeutic, the toxicity is significantly less because different path-ways will be targeted than those using monotherapeutic methods. Addi-tionally, combination therapy may be able to prevent toxic effects on normal cells while simultaneously producing cytotoxic effects on cancerous cells. This may occur if one drug in the combination regimen is antago-nistic, in terms of cytotoxicity, to another drug in normal cells, essentially protecting normal cells from cytotoxic effects. Nanocarrier-based drug delivery systems are a groundbreaking strategy that has changed the land-scape of therapy. The nanocarriers have been extensively applied for site-specific delivery and thus provided a useful delivery tool for combination therapy [80]. The rapid development of nanomedicine has allowed for the possibility of assembling several types of therapeutic agents into one nanocarrier through physical adsorption or chemical binding [81] to bring about multifunctional nanostructures for achieving the "mission" of com-bination therapy. Importantly, nanomaterials exhibit several advantages for use in combination cancer therapy. First, nanomaterials can passively accumulate at the tumor site via the enhanced permeability and retention effect [82,83]. Second, in the presence of diverse functional groups, the nanomaterial surfaces can be easily engineered with peptides, proteins, and other targeting agents to reduce uptake by the RES, and further, to specifically bind to tumor cell receptors for enhanced accumulation [84,85]. Third, the large surface-area-to-volume ratio of nanomaterials efficiently entraps high payloads of cargo and protects them from enzymatic degradation in complex physiological conditions [86]. Moreover, the functionalized nanomaterials can control the release of the loaded drugs via diverse stimuli (pH, light, etc.) [87,88]. In this regard, recent advances have turned to utilizing nanotechnology to design high-performance nanostructures through selective physical/chemical coloading of two or more kinds of therapeutic cargo for combination therapy, which may significantly improve therapeutic effectiveness and efficaciously treat those tumors that harbor resistance to monotherapy.

References

[1] Walker Jr B, Mouton CP. Nanotechnology and nanomedicine: a primer. J Natl Med Assoc 2006;98(12):1985.
[2] Olsson S, Lymberis A, Whitehouse D. European Commission activities in eHealth. Int J Circumpolar Health 2004;63(4):310−6.
[3] Feynman RP, Leighton RB, Sands M. The Feynman lectures on physics. The new millennium edition: mainly mechanics, radiation, and heat, vol. 1. Basic Books; 2011.
[4] Masciangioli T, Zhang W. Environmental technologies at the nanoscale. Environ Sci Technol 2003.
[5] Moghimi SM, Hunter AC, Murray JC. Nanomedicine: current status and future prospects. FASEB J 2005;19(3):311−30.
[6] Grodzinski P, et al. Integrating nanotechnology into cancer care. ACS Publications; 2019.
[7] Robinson JT, et al. Vertical nanowire electrode arrays as a scalable platform for intracellular interfacing to neuronal circuits. Nat Nanotechnol 2012;7(3):180−4.
[8] Wang J. Electrochemical glucose biosensors. Chem Rev 2008;108(2):814−25.
[9] Garimella R, Eltorai AE. Nanotechnology in orthopedics. J Orthop 2017;14(1):30−3.
[10] Zottel A, Videtič Paska A, Jovčevska I. Nanotechnology meets oncology: nanomaterials in brain cancer research, diagnosis and therapy. Materials 2019;12(10):1588.
[11] Zhao C-Y, et al. Nanotechnology for cancer therapy based on chemotherapy. Molecules 2018;23(4):826.
[12] Clauson RM, et al. Size-controlled iron oxide nanoplatforms with lipidoid-stabilized shells for efficient magnetic resonance imaging-trackable lymph node targeting and high-capacity biomolecule display. ACS Appl Mater Interfaces 2018;10(24):20281−95.
[13] Eide PK, et al. Magnetic resonance imaging provides evidence of glymphatic drainage from human brain to cervical lymph nodes. Sci Rep 2018;8(1):1−10.
[14] Skinder BM, Hamid S. Nanotechnology: a modern technique for pollution abatement. In: Bioremediation and biotechnology, vol. 4. Springer; 2020. p. 295−311.
[15] Ulijn RV, Jerala R. Peptide and protein nanotechnology into the 2020s: beyond biology. Chem Soc Rev 2018;47(10):3391−4.

[16] Bhatia S, et al. The challenges posed by cancer heterogeneity. Nat Biotechnol 2012;30(7):604—10.

[17] Dornan D, Settleman J. Dissecting cancer heterogeneity. Nat Biotechnol 2011;29(12):1095—6.

[18] Urbach D, et al. Cancer heterogeneity: origins and implications for genetic association studies. Trends Genet 2012;28(11):538—43.

[19] Smith BR, Gambhir SS. Nanomaterials for in vivo imaging. Chem Rev 2017;117(3):901—86.

[20] Huang H, Lovell JF. Advanced functional nanomaterials for theranostics. Adv Funct Mater 2017;27(2):1603524.

[21] Hong G, Antaris AL, Dai H. Near-infrared fluorophores for biomedical imaging. Nat Biomed Eng 2017;1(1):0010.

[22] Park S-M, et al. Towards clinically translatable in vivo nanodiagnostics. Nat Rev Mater 2017;2(5):17014.

[23] Chen H, et al. Rethinking cancer nanotheranostics. Nat Rev Mater 2017;2(7):17024.

[24] Li C. A targeted approach to cancer imaging and therapy. Nat Mater 2014;13(2):110.

[25] Piner RD, et al. "Dip-pen" nanolithography. Science 1999;283(5402):661—3.

[26] Canelas DA, Herlihy KP, DeSimone JM. Top-down particle fabrication: control of size and shape for diagnostic imaging and drug delivery. Wiley Interdiscip Rev Nanomed Nanobiotechnol 2009;1(4):391—404.

[27] Toy R, et al. The effects of particle size, density and shape on margination of nano-particles in microcirculation. Nanotechnology 2011;22(11):115101.

[28] Thompson AJ, Mastria EM, Eniola-Adefeso O. The margination propensity of ellip-soidal micro/nanoparticles to the endothelium in human blood flow. Biomaterials 2013;34(23):5863—71.

[29] Namdee K, et al. Margination propensity of vascular-targeted spheres from blood flow in a microfluidic model of human microvessels. Langmuir 2013;29(8):2530—5.

[30] Toy R, et al. Targeted nanotechnology for cancer imaging. Adv Drug Deliv Rev 2014;76:79—97.

[31] Fischer NO, et al. Colocalized delivery of adjuvant and antigen using nanolipoprotein particles enhances the immune response to recombinant antigens. J Am Chem Soc 2013;135(6):2044—7.

[32] Li H, et al. Alpha-alumina nanoparticles induce efficient autophagy-dependent cross-presentation and potent antitumour response. Nat Nanotechnol 2011;6(10):645—50.

[33] He Q, et al. Calcium phosphate nanoparticle adjuvant. Clin Diagn Lab Immunol 2000;7(6):899—903.

[34] Fernández A, et al. Inhibition of tumor-induced myeloid-derived suppressor cell function by a nanoparticulated adjuvant. J Immunol 2011;186(1):264—74.

[35] Moon JJ, et al. Interlayer-crosslinked multilamellar vesicles as synthetic vaccines for potent humoral and cellular immune responses. Nat Mater 2011;10(3):243—51.

[36] Li AV, et al. Generation of effector memory T cell—based mucosal and systemic immunity with pulmonary nanoparticle vaccination. Sci Transl Med 2013;5(204):204ra130.

[37] Kuai R, et al. Designer vaccine nanodiscs for personalized cancer immunotherapy. Nat Mater 2017;16(4):489—96.

[38] Kuai R, et al. Elimination of established tumors with nanodisc-based combination chemoimmunotherapy. Sci Adv 2018;4(4):eaao1736.

[39] Dziubla TD, Karim A, Muzykantov VR. Polymer nanocarriers protecting active enzyme cargo against proteolysis. J Contr Release 2005;102(2):427—39.

[40] Petrovsky N, Aguilar JC. Vaccine adjuvants: current state and future trends. Immunol Cell Biol 2004;82(5):488—96.

[41] Silva JM, et al. In vivo delivery of peptides and Toll-like receptor ligands by mannose-functionalized polymeric nanoparticles induces prophylactic and therapeutic anti-tumor immune responses in a melanoma model. J Contr Release 2015;198:91—103.

[42] Tam HH, et al. Sustained antigen availability during germinal center initiation enhances antibody responses to vaccination. Proc Natl Acad Sci USA 2016;113(43):E6639—48.

[43] Demento SL, et al. Role of sustained antigen release from nanoparticle vaccines in shaping the T cell memory phenotype. Biomaterials 2012;33(19):4957—64.

[44] Bachmann MF, Jennings GT. Vaccine delivery: a matter of size, geometry, kinetics and molecular patterns. Nat Rev Immunol 2010;10(11):787—96.

[45] Manolova V, et al. Nanoparticles target distinct dendritic cell populations according to their size. Eur J Immunol 2008;38(5):1404—13.

[46] Reddy S, et al. Nat Biotechnol 2007;25(1159):10.1038.

[47] Cho N-H, et al. A multifunctional core—shell nanoparticle for dendritic cell-based cancer immunotherapy. Nat Nanotechnol 2011;6(10):675—82.

[48] Embgenbroich M, Burgdorf S. Current concepts of antigen cross-presentation. Front Immunol 2018;9:1643.

[49] Shae D, et al. Endosomolytic polymersomes increase the activity of cyclic dinu-cleotide STING agonists to enhance cancer immunotherapy. Nat Nanotechnol 2019;14(3):269—78.

[50] Luo M, et al. A STING-activating nanovaccine for cancer immunotherapy. Nat Nanotechnol 2017;12(7):648.

[51] Bale SS, et al. Nanoparticle-mediated cytoplasmic delivery of proteins to target cellular machinery. ACS Nano 2010;4(3):1493—500.

[52] Wilson B, et al. Nanomed Nanotechnol Biol Med 2010;6(1):144—52.

[53] Singhvi G, Banerjee S, Khosa A. Lyotropic liquid crystal nanoparticles: a novel improved lipidic drug delivery system. In: Organic materials as smart nanocarriers for drug delivery. Elsevier; 2018. p. 471—517.

[54] Khan DR. The use of nanocarriers for drug delivery in cancer therapy. J Canc Sci Ther 2010;2(3):58—62.

[55] Khodabandehloo H, Zahednasab H, Hafez AA. Nanocarriers usage for drug delivery in cancer therapy. Iran J Cancer Prev 2016;9(2).

[56] Shi D, Bedford NM, Cho HS. Engineered multifunctional nanocarriers for cancer diagnosis and therapeutics. Small 2011;7(18):2549—67.

[57] Torchilin VP. Targeted pharmaceutical nanocarriers for cancer therapy and imaging. AAPS J 2007;9(2):E128—47.

[58] Xiong X-B, Lavasanifar A. Traceable multifunctional micellar nanocarriers for cancer-targeted co-delivery of MDR-1 siRNA and doxorubicin. ACS Nano 2011;5(6):5202—13.

[59] Crucho CI, Barros MT. Polymeric nanoparticles: a study on the preparation variables and characterization methods. Mater Sci Eng C 2017;80:771—84.

[60] Huynh NT, et al. The rise and rise of stealth nanocarriers for cancer therapy: passive versus active targeting. Nanomedicine 2010;5(9):1415—33.

[61] Pasut G, Veronese FM. PEG conjugates in clinical development or use as anticancer agents: an overview. Adv Drug Deliv Rev 2009;61(13):1177—88.

[62] Kamaly N, et al. Degradable controlled-release polymers and polymeric nanoparticles: mechanisms of controlling drug release. Chem Rev 2016;116(4):2602—63.

[63] Alexis F, et al. New frontiers in nanotechnology for cancer treatment. In: Urologic oncology: seminars and original investigations. Elsevier; 2008.

[64] Muhamad N, Plengsuriyakarn T, Na-Bangchang K. Application of active targeting nanoparticle delivery system for chemotherapeutic drugs and traditional/herbal med-icines in cancer therapy: a systematic review. Int J Nanomed 2018;13:3921.

[65] Lian T, Ho RJ. Trends and developments in liposome drug delivery systems. J Pharmaceut Sci 2001;90(6):667—80.

[66] Jaracz S, et al. Recent advances in tumor-targeting anticancer drug conjugates. Bioorg Med Chem 2005;13(17):5043—54.

[67] Bulbake U, et al. Liposomal formulations in clinical use: an updated review. Pharmaceutics 2017;9(2):12.

[68] Reddy JA, Low PS. Enhanced folate receptor mediated gene therapy using a novel pH-sensitive lipid formulation. J Contr Release 2000;64(1—3):27—37.

[69] Slepushkin VA, et al. Sterically stabilized pH-sensitive liposomes intracellular delivery of aqueous contents and prolonged circulation in vivo. J Biol Chem 1997;272(4):2382—8.

[70] Turk MJ, et al. Characterization of a novel pH-sensitive peptide that enhances drug release from folate-targeted liposomes at endosomal pHs. Biochim Biophys Acta Biomembr 2002;1559(1):56—68.

[71] Sun C, Lee JS, Zhang M. Magnetic nanoparticles in MR imaging and drug delivery. Adv Drug Deliv Rev 2008;60(11):1252—65.

[72] Lin J-J, et al. Folic acid—Pluronic F127 magnetic nanoparticle clusters for combined targeting, diagnosis, and therapy applications. Biomaterials 2009;30(28):5114—24.

[73] Xie J, et al. Surface-engineered magnetic nanoparticle platforms for cancer imaging and therapy. Acc Chem Res 2011;44(10):883—92.

[74] Mornet S, et al. Magnetic nanoparticle design for medical diagnosis and therapy. J Mater Chem 2004;14(14):2161—75.

[75] Veiseh O, Gunn JW, Zhang M. Design and fabrication of magnetic nanoparticles for targeted drug delivery and imaging. Adv Drug Deliv Rev 2010;62(3):284—304.

[76] Brunetaud J, et al. Non-PDT uses of lasers in oncology. Laser Med Sci 1995;10(1):3—8.

[77] Huang X, et al. Plasmonic photothermal therapy (PPTT) using gold nanoparticles. Laser Med Sci 2008;23(3):217.

[78] Yap TA, Omlin A, De Bono JS. Development of therapeutic combinations targeting major cancer signaling pathways. J Clin Oncol 2013;31(12):1592—605.

[79] Partridge AH, Burstein HJ, Winer EP. Side effects of chemotherapy and combined chemohormonal therapy in women with early-stage breast cancer. JNCI Monogr 2001;2001(30):135—42.

[80] Gao Z, Zhang L, Sun Y. Nanotechnology applied to overcome tumor drug resistance. J Contr Release 2012;162(1):45—55.

[81] Xie J, Lee S, Chen X. Nanoparticle-based theranostic agents. Adv Drug Deliv Rev 2010;62(11):1064—79.

[82] Perry JL, et al. Mediating passive tumor accumulation through particle size, tumor type, and location. Nano Lett 2017;17(5):2879—86.

[83] Maeda H. Toward a full understanding of the EPR effect in primary and metastatic tumors as well as issues related to its heterogeneity. Adv Drug Deliv Rev 2015;91:3—6.

[84] Rao L, et al. Cancer cell membrane-coated upconversion nanoprobes for highly specific tumor imaging. Adv Mater 2016;28(18):3460—6.

[85] Chen F, Cai W. Tumor vasculature targeting: a generally applicable approach for functionalized nanomaterials. Small 2014;10(10):1887—93.

[86] Biju V. Chemical modifications and bioconjugate reactions of nanomaterials for sensing, imaging, drug delivery and therapy. Chem Soc Rev 2014;43(3):744—64.

[87] Tseng Y-J, et al. A versatile theranostic delivery platform integrating magnetic resonance imaging/computed tomography, pH/cis-diol controlled release, and targeted therapy. ACS Nano 2016;10(6):5809—22.

[88] Karimi M, et al. Smart micro/nanoparticles in stimulus-responsive drug/gene delivery systems. Chem Soc Rev 2016;45(5):1457—501.

CHAPTER 3

Passive targeting

Passive targeting exploits the accumulation of nanocarriers at sites of interest, such as tumor locations. Convection (a passive-diffusion process) is mediated by nanocarrier transportation via pores in leaky capillaries that are present in tumor masses and tissues that trigger angiogenesis. This process occurs in conjunction with the enhanced permeability and retention (EPR) effect. However, passive targeting is not classified as a type of selective targeting. The EPR effect applies not only to tumors but also to off-target organs such as the spleen, liver, and lungs. Passive targeting with the EPR effect can be used to overcome dilemmas affecting nearly every kind of tumor targeting currently in use for deficiency in tumor selectivity. However, referring to this as "selectivity" might be seen as deceptive. Although nanostructures (NSs) are administered into the blood circulatory system, there is no "selectivity" with reference to the EPR effect. Nanostructures are distributed throughout the body with the aim of "passive targeting" via the EPR effect to attain inconsistent distribution of NSs by concentrating on the tumors. In fact, in addition to the EPR effect, passive targeting of cancers is also conducted through the tumor microenvironment (TME) and direct local delivery.

1. Enhanced permeability and retention effect

As the size of a tumor approaches $2 \, \text{mm}^3$, the delivery of nutrients and oxygen by simple diffusion becomes insufficient. Therefore, the formation of new blood vessels in the TME is essential to support the rapid growth of malignant tumors [1]. Hypoxia occurs in developing tumors because the lack of blood supply triggers the release of angiogenic growth factors from the neoplastic tissue, thus enabling additional tumor growth. The imbalance between angiogenic growth factors and matrix metalloproteinases (MMPs) in neoplastic tissues results in pronounced vessel disorganization associated with the formation of highly porous large-gap junctions between endothelial cells (ECs). Overall, defective basement membranes and incomplete

Targeted Cancer Imaging
ISBN 978-0-12-824513-2
https://doi.org/10.1016/B978-0-12-824513-2.00001-2
37

coverage with perivascular cells promote leakage and accumulation of agents administered into tumor tissues [2−4].

The smooth muscle layer surrounding ECs in normal blood vessels does not exist in tumor blood vessels [5]. Normal vasculature typically possesses tight, impermeable EC junctions. Molecules with a diameter greater than 2−4 nm cannot pass through the junctions of normal vessels. Hence, larger NSs are generally excluded from normal tissues and organs. Neoplastic tissues with leaky vasculature permit entry of macromolecules up to 600 nm diameter into the tissue [3,6]. Moreover, the defective lymphatic drainage system of tumor tissues enables NSs to remain in neoplastic tissue for a prolonged period [7]. The EPR effect plays a key role in the delivery of targeted agents to the TME [8,9] (Fig. 3.1).

The EPR effect was first observed in living animals using the fluorescent dye "Evans blue." After injection of the dye, the tumor mass was selectively stained blue [10]. The EPR effect is highly variable and differs according to tumor type, mass, and size as well as the penetration and accumulation locations within the tumor mass. Furthermore, the mononuclear phagocyte system and tumor-associated immune cell activity modulate the circulation profile of the targeted agent and tumor transport, accumulation, and release.

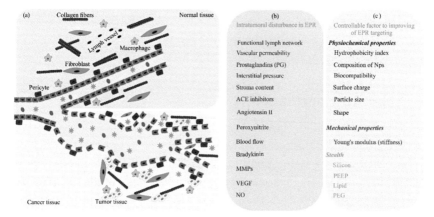

Figure 3.1 (A) Schematic illustration of the enhanced permeability and retention (EPR) effect in cancer and normal tissue. (B) Intratumoral disturbance parameter in cancerous tissue. (C) Controllable factor to improve EPR effect targeting—"the stealth design of NPs aims to have maximum circulation half-life ensured of presenting continuous delivery into the tumor site with the leaky vasculature." *ACE*, angiotensin-converting-enzyme; *MMPs*, matrix metalloproteinases; *NO*, nitric oxide; *PEEP*, poly(ethyl ethylene phosphate); *PEG*, polyethylene glycol; *VEGF*, vascular endothelial growth factors.

The pore size of the leaky vasculature varies significantly across different tumor types. Pore size plays a vital role in the tumor accumulation of targeted agents. Because pore sizes vary so much, designing a particle with an optimal shape and size is extremely difficult. For example, one study compared rhodamine B with MW 479 Da and tetramethylrhodamine isothiocyanate conjugated-bovine serum albumin (TRITC–BSA) with MW 67000 Da (Fig. 3.2). The observation was that rhodamine B did not

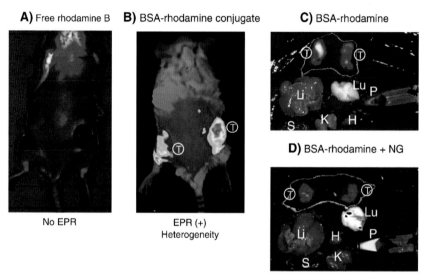

Figure 3.2 *Fluorescence imaging based on the enhanced permeability and retention (EPR) effect.* The EPR effect-based uptake of a fluorescent imaging nanoprobe in the tumor was compared with uptake of the parental low-molecular-weight (LMW) fluorescent probe in vivo. (A) 24 h after intravenous injection of the LMW fluorescent probe of rhodamine B into S-180 tumor-bearing mice, no distinct tumor was visible. (B) Injection of tetramethylrhodamine isothiocyanate conjugated-bovine serum albumin (TRITC-BSA) (67 kDa) resulted in highly tumor-selective fluorescence under the same experimental conditions. (C) At 24 h, S-180 tumor-bearing mice were dissected, and each organ was imaged with an IVIS imaging system. Results showed that only the tumor tissues exhibited significant fluorescence. (D) Same as C, except that nitroglycerin (NG) ointment was applied to the skin, and then the EPR effect and tumor targeting were evaluated. In D, the cut surfaces of tumor tissues showed more homogeneous tumor uptake of the TRITC-BSA probe, and more TRITC-BSA remained in the blood, indicating that the EPR effect depends on time and would progressively increase. In C and D, fluorescence is seen only in tumor tissue. *H*, heart; *Li*, liver; *Lu*, lung; *P*, plasma; *S*, spleen; *T*, tumor. *(Reproduced with permission from Maeda H, Nakamura H, Fang J. The EPR effect for macromolecular drug delivery to solid tumors: improvement of tumor uptake, lowering of systemic toxicity, and distinct tumor imaging in vivo. Adv Drug Deliv Rev 2013;65(1):71–9 [Copyright 2013, Elsevier]).*

A) 2 h   **B)** 24 h   **C)** 48 h

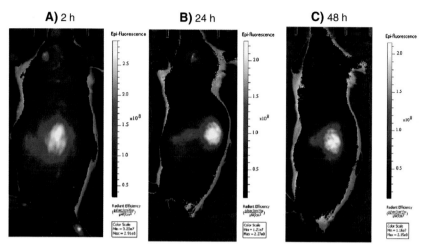

Figure 3.3 *Tumor imaging with indocyanine green.* Indocyanine green (ICG) was injected intravenously into S-180 tumor-bearing mice, and fluorescent imaging was carried out at 2, 24, and 48 h by IVIS imaging system. The ICG bound to albumin and behaved as a macromolecule. The images show the increase in tumor contrast over time. That is, nonspecific delivery of the agent to normal tissues was cleared via the lymphatic system, thus improving the contrast of the tumor image (2 vs. 48 h). *(Reproduced with permission from Maeda H, Nakamura H, Fang J. The EPR effect for macromolecular drug delivery to solid tumors: improvement of tumor uptake, lowering of systemic toxicity, and distinct tumor imaging in vivo. Adv Drug Deliv Rev 2013;65(1):71—9 [Copyright 2013, Elsevier]).*

provide appreciable fluorescence in the tumor mass. However, TRITC-BSA emitted considerable fluorescence that remained in the tumor even 72 h after injection [11]. Another study used indocyanine green (ICG) to evaluate hepatic function in healthy mice [12]. ICG initially bound to albumin and globulin was released as a free dye until it cleared the plasma (half-life <20 min) (Fig. 3.3). In contrast, albumin bound to ICG accumulated specifically in the tumor mass in mice. Over time, ICG accumulation steadily increased. This effect was not seen in normal tissue because the lymphatic system had cleared the dye from the bloodstream. Therefore, as the elapsed time increased, the image resolution increased [11].

The EPR effect also depends on the type, size, location, and total blood volume of the tumor. Blood volume is especially important because it influences biodistribution. In addition, increased blood volume is accompanied by increases in tumor size and consequent reductions in tumor uptake of the nanoplatform. The EPR effect in humans has been observed mostly in squamous cell carcinoma of the head and neck and only rarely in cancers of the lung and breast [13]. Evaluation of the accumulation of various NP sizes in four different subcutaneous flank tumor models found that each tumor possessed unique accumulation properties. Another factor directly affecting tumor accumulation is the density of microvessels within the tumor. The density of microvessels varies by tumor type [14].

To develop more efficient targeting methods that take advantage of the EPR effect, recent studies have focused on three areas: (1) modification of the EPR effect using antiangiogenic agents; (2) reduction in interstitial fluid pressure; and (3) application of external stimuli (i.e., ultrasound and temperature) to increase tumor permeability [15].

In recent decades, preclinical research into EPR effects on tumor-targeted therapy has increased; nevertheless, expected clinical translation has been challenged by differences between mice tumor models and human tumors in terms of (1) rate of development, metabolic rate, host life span, and size relative to host, (2) tumor growth rate, which is faster in rodents, (3) the large tumor-to-body weight ratio in mice compared with humans, which significantly alters the pharmacokinetic effects of nanocarriers, and (4) the heterogeneity of EPR, which affects TME-associated endothelium structure, the rate of blood flow within the tumor region, pericyte coverage, and extracellular matrix density [16,17].

2. Tumor microenvironment

The TME is another passive targeting strategy—it uses distinctive tumor environment conditions for tumor-activated therapy of prodrugs. Rapidly growing and hyperproliferative cancer cells exhibit a developed metabolic rate; however, oxygen and nutrient delivery is generally not sufficient to support this. Consequently, cancerous cells make use of glycolysis to obtain additional energy, resulting in an acidic milieu [18]. Furthermore, cancer cells articulate and discharge distinctive enzymes, such as MMPs, involved in their motion and in endurance mechanisms [19]. The effective agents are coupled with tumor-specific particles and stay inert, waiting to arrive at the target. The therapeutic agent is coupled with a biocompatible polymer through an ester linker group. The association agents are hydrolyzed by a cancer-specific enzyme or through changes in pH (higher or lower) at the site of the tumor, at which point the NS delivers the cargo [20]. For example, an albumin-bound doxorubicin (DOX) that integrated an MMP-2 specific octapeptide series between the anticancer drug and the nanocarrier was reported to be proficiently and specially cleaved by MMP-2 in an in vitro examination [21].

3. Direct local delivery

Another passive targeting strategy is straight local release of therapeutic agents to the tumor tissues. This technique has the comprehensible benefit of excluding therapeutic agents from systemic circulation. However, administration can be quite intrusive because it entails injections or surgical procedures. For a few cancers that are difficult to access (such as lung cancers), employing this approach is nearly impossible [22].

4. Nanostructure delivery system for cancer with passive targeting

Several passively targeted NSs are in clinical use, such as Genexol-PM in Korea and ProLindac in the United States [23,24]. Several additional nanosystems (AZD2811, CPX-1, and NK911) have confirmed safety and therapeutic effectiveness in clinical studies [25–27]. Starting from a considerable number of nanocarriers, only a few nanomedicines have been accepted for use in cancer targeting. Even though most candidate nano-systems change pharmacokinetics, toxicological profiles, and drug solubility, some have also demonstrated notable endurance advantages and improved therapeutic efficiency compared with the parent medicine in clinical studies. Abraxane (albumin-bound paclitaxel nanoparticles) is one example—it established noticeably high response rates compared with standard paclitaxel in a phase 3 clinical trial of patients with metastatic cancer [28]. Likewise, FDA-approved Vyxeos (CPX-351), a liposomal preparation of cytarabine and daunorubicin, demonstrated enhanced endurance of 9.56 months compared with 5.95 months for daunorubicin and cytarabine delivered in their free forms in patients identified with vulnerable acute myeloid leukemia [29]. Our understanding of EPR efficiency is controlled by the limited information gained from preclinical cancer models to precisely categorize solid tumors in individuals. In fact, the most frequently used subcutaneous xenograft tumors are quickly expanding, resulting in extremely high-EPR cancers that might provide incorrect assessments of the therapeutic advantages of nanosystems in treatments that depend on EPR effect-based passive targeting. Moreover, there is limited patient-based investigational information on the EPR effect itself in addition to its effects on the buildup of an effective cargo in the tumor site, which can be understood as clinical effectiveness [30,31]. Other research on the EPR effect in various tumors, as well as improvements in preclinical models, is therefore necessary for the design of nanocarriers with enhanced tumor penetration and therapeutic effects [31,32]. The link between tumor vascularization and EPR effect-based passive targeting was explored by the Theek group in a subcutaneous tumor model [33]. Using both contrast-enhanced computed tomography fluorescence molecular and ultrasound imaging systems, they established heterogeneous buildup of near infrared-labeled polymeric nanocarriers (pHPMA-Dy750) among and inside the tumors (5%−12%). In another study, copper-loaded polyethylene glycol (PEG)ylated nanoliposome was investigated by the Hansen group. Moreover, the EPR effect of PEGylated liposomes was evaluated with micro-PET/CT imaging modalities [13]. Evaluation of 11 dogs bearing spontaneous tumors revealed that EPR is a predominant feature in a few tumors and results in greater liposome accumulation; however, this might not be so widespread that it affects every solid tumor. FDA-approved ferumoxytol (carboxymethyl dextran-coated magnetic NPs) could be used as a substitute or companion element for the intratumoral transfer, distribution, and pharmacokinetics of a therapeutic nanocarrier rooted in poly(D, L-lactic-co-glycolic acid)-b-poly(ethylene glycol), as demonstrated by Miller et al. [34]. Lee et al. developed ^{64}Cu-labeled HER2-targeted liposomes along with PET/CT to measure the accumulation of cargo in patients with HER2-positive metastatic breast cancer [35]. The maximum accumulation of liposomes occurred 24−48 h after treatment; patients were categorized according to liposomal abrasion deposition by a cut point similar to a reaction threshold as strongly believed in preclinical investigations. Patients labeled with elevated liposomal abrasion deposition were associated with more positive therapy results. These results reveal that the use of imaging modalities for the assessment and characterization of EPR may ultimately allow clinicians to preselect patients with higher-EPR cancers who are expected to respond to passively targeted nanocarriers with enhanced therapeutic effects. A meta-analysis of preclinical data based on a nanocarrier delivery system for cancer reported during the past decade found a mean of nearly 0.7% for the injected dose of nanocarriers that arrives at target tumors [36]. This amount appears to be very small, elevating concerns about the suitability of the EPR effect and the management of low-EPR cancers. However, a delivery effectiveness of nearly 0.7% for nanocarriers is considerably better than the delivery effectiveness of most chemotherapeutic preparations commonly used in hospitals, including paclitaxel, docetaxel, and DOX. Van Vlerken et al. established a delivery effectiveness of 0.6% of the injected dose for paclitaxel-loaded nanocarriers in comparison with 0.2% of the injected dose for free paclitaxel in a preclinical investigation [37—39]. These results are hopeful and indicate the benefits of nanocarriers for tumor-targeted delivery of cargo. Nevertheless, the delivery effectiveness of nanocarriers could be enhanced to better use their therapeutic advantages. Enhancing EPR effects with angiotensin II-induced heat-based or hypertension vasodilation might be an answer; however, such a system could make clinical transformation of nanocarriers difficult. A more promising and comparatively adaptable solution, particularly for low-EPR tumors, is gracefully engineered delivery methods that make use of non-EPR advances for tumor targets. For example, Xu and coworkers developed injectable nanoparticle generators (iNPGs) that tackle many physiological barriers [40]. The iNPG is a discoidal micrometer-sized silicon particle that is nanoporous and can be burdened with drug polymer conjugates, thereby controlling tumor growth values because of normal tropism along with improved vascular dynamics. The iNPG delivers polymeric cargo by self-assembling NSs that transfer to the perinuclear area and thereby bypass the drug efflux pump. Greater efficiency was observed with iNPGs in 4T1 mouse and MDA-MB-231 models of metastatic breast cancer compared with their individual components as well as in other existing therapeutic dosage systems. The delivery effectiveness of nano-carriers could be significantly enhanced by such realistically engineered nanosystems. Cell-mediated delivery of nanocarriers may be an additional EPR-free advance to improve tumor cell targeting in low-EPR tumors or certain metastatic tumor sites that cannot be reached via passive targeting. This method uses the suitability of definite cell type to house or travel to such tumors [41]. Coating nanocarriers can help nanocarrier—protein interactions in the bloodstream [42,43]. Targeting strategies are used to ensure that an adequate number of nanosystems reach tumor cells. In a passive targeting approach, the individual receives the benefit of the elevated endocytic uptake of cancer cells as well as the penetrable vasculature in the tumor site, which allows for high uptake of nanosystems compared with healthy tissues [44].

In the most passive targeting nanosystems, surface coating with PEG is accomplished for biocompatibility and "stealth" purposes [32,45,46]. Significantly, improved nanosystem surface hydrophilicity can prevent its uptake by cancer cells, thus hindering the qualified delivery of a drug to tumor sites with passive targeting nanosystems [32,47,48]. Nevertheless, PEG-based block copolymers have been used in many passive targeting polymeric nanocarriers, including NK911, SP1049C, and Genexol-PM. Among them, SP1049C is a pluronic-based polymeric micellar nanocarrier for DOX. At present, it is being investigated in phase 2 clinical trials for metastatic cancers of the esophagus versus the usual chemotherapeutic protocols [49]. Another polymeric micellar nanocarrier that acts through a passive targeting mechanism is NK911, containing PEG, DOX, and polyaspartic acid, which is currently being investigated in phase 2 clinical trials for a variety of cancers [49]. CRLX101, a camptothecin-cyclodextrin polymeric conjugate, has shown improved pharmacokinetic effectiveness in clinical investigations [50].

References

[1] Bergers G, Benjamin LE. Tumorigenesis and the angiogenic switch. Nat Rev Cancer 2003;3(6):401—10.
[2] Greish K. Enhanced permeability and retention of macromolecular drugs in solid tumors: a royal gate for targeted anticancer nanomedicines. J Drug Target 2007;15(7—8):457—64.
[3] Cho K, Wang X, Nie S, Shin DM. Therapeutic nanoparticles for drug delivery in cancer. Clin Cancer Res 2008;14(5):1310—6.
[4] Iyer AK, Khaled G, Fang J, Maeda H. Exploiting the enhanced permeability and retention effect for tumor targeting. Drug Discov Today 2006;11(17—18):812—8.
[5] Skinner SA, Tutton PJ, O'Brien PE. Microvascular architecture of experimental colon tumors in the rat. Cancer Res 1990;50(8):2411—7.
[6] Talekar M, Kendall J, Denny W, Garg S. Targeting of nanoparticles in cancer: drug delivery and diagnostics. Anti Cancer Drugs 2011;22(10):949—62.
[7] Weissleder R, Nahrendorf M, Pittet MJ. Imaging macrophages with nanoparticles. Nat Mater 2014;13(2):125.
[8] Matsumura Y, Maeda H. A new concept for macromolecular therapeutics in cancer chemotherapy: mechanism of tumoritropic accumulation of proteins and the antitumor agent smancs. Cancer Res 1986;46(12 Part 1):6387—92.
[9] Maeda H, Matsumura Y. Tumoritropic and lymphotropic principles of macromolecular drugs. Crit Rev Ther Drug Carrier Syst 1989;6(3):193—210.
[10] Xiao K, Luo J, Li Y, Xiao W, Lee JS, Gonik AM, et al. The passive targeting of polymeric micelles in various types and sizes of tumor models. Nanosci Nanotechnol Lett 2010;2(2):79—85.
[11] Maeda H, Nakamura H, Fang J. The EPR effect for macromolecular drug delivery to solid tumors: improvement of tumor uptake, lowering of systemic toxicity, and distinct tumor imaging in vivo. Adv Drug Deliv Rev 2013;65(1):71—9.
[12] Nakamura H, Liao L, Hitaka Y, Tsukigawa K, Subr V, Fang J, et al. Micelles of zinc protoporphyrin conjugated to N-(2-hydroxypropyl) methacrylamide (HPMA) copolymer for imaging and light-induced antitumor effects in vivo. J Contr Release 2013;165(3):191—8.
[13] Hansen AE, Petersen AL, Henriksen JR, Boerresen B, Rasmussen P, Elema DR, et al. Positron emission tomography based elucidation of the enhanced permeability and

retention effect in dogs with cancer using copper-64 liposomes. ACS Nano 2015;9(7):6985−95.

[14] Perry JL, Reuter KG, Luft JC, Pecot CV, Zamboni W, DeSimone JM. Mediating passive tumor accumulation through particle size, tumor type, and location. Nano Lett 2017;17(5):2879−86.

[15] Nichols JW, Bae YH. EPR: evidence and fallacy. J Contr Release 2014;190:451−64.

[16] Gillies RJ, Schomack PA, Secomb TW, Raghunand N. Causes and effects of heterogeneous perfusion in tumors. Neoplasia 1999;1(3):197−207.

[17] Danhier F. To exploit the tumor microenvironment: since the EPR effect fails in the clinic, what is the future of nanomedicine? J Contr Release 2016;244:108−21.

[18] Pelicano H, Martin D, Xu R, Huang P. Glycolysis inhibition for anticancer treatment. Oncogene 2006;25(34):4633−46.

[19] Deryugina EI, Quigley JP. Matrix metalloproteinases and tumor metastasis. Cancer Metastasis Rev 2006;25(1):9−34.

[20] Chari RV. Targeted delivery of chemotherapeutics: tumor-activated prodrug therapy. Adv Drug Deliv Rev 1998;31(1−2):89−104.

[21] Mansour AM, Drevs J, Esser N, Hamada FM, Badary OA, Unger C, et al. A new approach for the treatment of malignant melanoma: enhanced antitumor efficacy of an albumin-binding doxorubicin prodrug that is cleaved by matrix metalloproteinase 2. Cancer Res 2003;63(14):4062−6.

[22] Kim GJ, Nie S. Targeted cancer nanotherapy. Mater Today 2005;8(8):28−33.

[23] Shi J, Kantoff PW, Wooster R, Farokhzad OC. Cancer nanomedicine: progress, challenges and opportunities. Nat Rev Cancer 2017;17(1):20.

[24] Sanna V, Pala N, Sechi M. Targeted therapy using nanotechnology: focus on cancer. Int J Nanomed 2014;9:467.

[25] Awada A, Bondarenko I, Bonneterre J, Nowara E, Ferrero J, Bakshi A, et al. A randomized controlled phase II trial of a novel composition of paclitaxel embedded into neutral and cationic lipids targeting tumor endothelial cells in advanced triple-negative breast cancer (TNBC). Ann Oncol 2014;25(4):824−31.

[26] Burris HA, Wang JS-Z, Johnson ML, Falchook GS, Jones SF, Strickland DK, et al. A phase I, open-label, first-time-in-patient dose escalation and expansion study to assess the safety, tolerability, and pharmacokinetics of nanoparticle encapsulated Aurora B kinase inhibitor AZD2811 in patients with advanced solid tumours. Am Soc Clin Oncol 2017.

[27] Batist G, Sawyer M, Gabrail N, Christiansen N, Marshall J, Spigel D, et al. A multicenter, phase II study of CPX-1 liposome injection in patients (pts) with advanced colorectal cancer (CRC). J Clin Oncol 2008;26(15_Suppl. l):4108.

[28] Gradishar WJ, Tjulandin S, Davidson N, Shaw H, Desai N, Bhar P, et al. Phase III trial of nanoparticle albumin-bound paclitaxel compared with polyethylated castor oil−based paclitaxel in women with breast cancer. J Clin Oncol 2005;23(31):7794−803.

[29] Lancet JE, Uy GL, Cortes JE, Newell LF, Lin TL, Ritchie EK, et al. Final results of a phase III randomized trial of CPX-351 versus 7+ 3 in older patients with newly diagnosed high risk (secondary) AML. Am Soc Clin Oncol 2016.

[30] Bogart LK, Pourroy G, Murphy CJ, Puntes V, Pellegrino T, Rosenblum D, et al. Nanoparticles for imaging, sensing, and therapeutic intervention. ACS Nano 2014.

[31] Byrne JD, Betancourt T, Brannon-Peppas L. Active targeting schemes for nanoparticle systems in cancer therapeutics. Adv Drug Deliv Rev 2008;60(15):1615−26.

[32] Alexis F, Pridgen E, Molnar LK, Farokhzad OC. Factors affecting the clearance and biodistribution of polymeric nanoparticles. Mol Pharm 2008;5(4):505−15.

[33] Theek B, Gremse F, Kunjachan S, Fokong S, Pola R, Pechar M, et al. Characterizing EPR-mediated passive drug targeting using contrast-enhanced functional ultrasound imaging. J Contr Release 2014;182:83−9.

[34] Miller MA, Gadde S, Pfirschke C, Engblom C, Sprachman MM, Kohler RH, et al. Predicting therapeutic nanomedicine efficacy using a companion magnetic resonance imaging nanoparticle. Sci Transl Med 2015;7(314):314ra183.

[35] Lee H, Shields AF, Siegel BA, Miller KD, Krop I, Ma CX, et al. ^{64}Cu-MM-302 positron emission tomography quantifies variability of enhanced permeability and retention of nanoparticles in relation to treatment response in patients with metastatic breast cancer. Clin Cancer Res 2017;23(15):4190−202.

[36] Wilhelm S, Tavares AJ, Dai Q, Ohta S, Audet J, Dvorak HF, et al. Analysis of nanoparticle delivery to tumours. Nat Rev Mater 2016;1(5):1−12.

[37] van Vlerken LE, Duan Z, Little SR, Seiden MV, Amiji MM. Biodistribution and pharmacokinetic analysis of paclitaxel and ceramide administered in multifunctional polymer-blend nanoparticles in drug resistant breast cancer model. Mol Pharm 2008;5(4):516−26.

[38] Cui Y, Zhang M, Zeng F, Jin H, Xu Q, Huang Y. Dual-targeting magnetic PLGA nanoparticles for codelivery of paclitaxel and curcumin for brain tumor therapy. ACS Appl Mater Interfaces 2016;8(47):32159−69.

[39] Shi J, Xiao Z, Kamaly N, Farokhzad OC. Self-assembled targeted nanoparticles: evolution of technologies and bench to bedside translation. Acc Chem Res 2011;44(10):1123−34.

[40] Xu R, Zhang G, Mai J, Deng X, Segura-Ibarra V, Wu S, et al. An injectable nanoparticle generator enhances delivery of cancer therapeutics. Nat Biotechnol 2016;34(4):414−8.

[41] Levy O, Brennen WN, Han E, Rosen DM, Musabeyezu J, Safaee H, et al. A prodrug-doped cellular Trojan Horse for the potential treatment of prostate cancer. Biomaterials 2016;91:140−50.

[42] Monopoli MP, Åberg C, Salvati A, Dawson KA. Biomolecular coronas provide the biological identity of nanosized materials. Nat Nanotechnol 2012;7(12):779−86.

[43] Krpetić Ž, Anguissola S, Garry D, Kelly PM, Dawson KA. Nanomaterials: impact on cells and cell organelles. Nanomaterial 2014:135−56.

[44] Barreto JA, O'Malley W, Kubeil M, Graham B, Stephan H, Spiccia L. Nanomaterials: applications in cancer imaging and therapy. Adv Mater 2011;23(12):H18−40.

[45] Otsuka H, Nagasaki Y, Kataoka K. PEGylated nanoparticles for biological and pharmaceutical applications. Adv Drug Deliv Rev 2003;55(3):403−19.

[46] Avgoustakis K. Pegylated poly (lactide) and poly (lactide-co-glycolide) nanoparticles: preparation, properties and possible applications in drug delivery. Curr Drug Deliv 2004;1(4):321−33.

[47] Knop K, Hoogenboom R, Fischer D, Schubert US. Poly (ethylene glycol) in drug delivery: pros and cons as well as potential alternatives. Angew Chem Int Ed 2010;49(36):6288−308.

[48] Hatakeyama H, Akita H, Harashima H. A multifunctional envelope type nano device (MEND) for gene delivery to tumours based on the EPR effect: a strategy for overcoming the PEG dilemma. Adv Drug Deliv Rev 2011;63(3):152−60.

[49] Valle JW, Armstrong A, Newman C, Alakhov V, Pietrzynski G, Brewer J, et al. A phase 2 study of SP1049C, doxorubicin in P-glycoprotein-targeting pluronics, in patients with advanced adenocarcinoma of the esophagus and gastroesophageal junction. Invest New Drugs 2011;29(5):1029−37.

[50] Young C, Schluep T, Hwang J, Eliasof S. CRLX101 (formerly IT-101) a novel nanopharmaceutical of camptothecin in clinical development. Curr Bioact Compd 2011;7(1):8−14.

CHAPTER 4

Tumor-specific imaging probes in preclinical applications and clinical trials

1. Active targeting

Active targeting agents can selectively transport nanoparticles (NPs) into the tumor mass and bind with high affinity to molecules expressed on cancer cell surfaces, leading to endocytosis-mediated cell uptake [1]. A surface-functionalized nanocarrier using specific ligands complements the passive targeting approach to improve nanocarrier delivery and tumor localization [2].

2. General approach to active targeting

Conventional active targeting methods consist of targeting molecules or receptors overexpressed in selected types of cancers [3]. They can be categorized into three subsets: (1) targetable factors in the tumor micro-environment (TME), such as hypoxia, pH, matrix metalloproteinase (MMP) enzymes, and fibronectin; (2) targetable molecules on the surface of tumor endothelial cells (ECs), such as vascular endothelial growth factor (VEGF) receptors, integrins, and vascular cell-adhesion molecule-1 (VCAM-1); and (3) targetable molecules on or within the cancer cells themselves, such as transferrin receptor, folate receptor, epidermal growth factor receptor (EGFR), glucose transporter, and cathepsins [4].

3. Tumor microenvironment targeting

3.1 Hypoxia targeting

The imbalance between blood vessel formation and the rate of tumor cell proliferation causes hypoxia due to oxygen-deprived conditions. Hypoxia leads to upregulation of hypoxia-inducible factor 1α and triggers the unfolded protein response. Both transcription elements influence the

Targeted Cancer Imaging
ISBN 978-0-12-824513-2
https://doi.org/10.1016/B978-0-12-824513-2.00004-8

expression of genes associated with tumor initiation, progression, malignancy, metastasis, and drug resistance [5] (Fig. 4.1). The decreased O_2 concentration causes the accumulation of reduced nicotinamide adenine dinucleotide (NADH) and flavin adenine dinucleotide ($FADH_2$) species in the TME, thus further reducing the remaining oxygen and the production of reactive oxygen species (ROS) [6].

The presence of hypoxia indicates that cancer cells possess an insufficient blood supply. As a result, only a small portion of drugs and contrast agents can be delivered to the cancer site [7]. The redox balance and oxygen concentration are two intrinsic properties available for hypoxia targeting [8]. Different techniques are required to monitor the degree of hypoxia and to directly measure pO_2 in the cells. The redox balance is influenced mainly by redox agents, such as cysteine and glutathione. The degree of hypoxia in cancerous tissue can be deduced by measuring the local concentrations of reducing enzymes such as azoreductase and nitroreductase [9]. Chemical groups such as nitro, azo, and quinone groups can be used to target hypoxia and for reversible sensing between normoxia/hypoxia.

Hypoxia occurs in solid tumors where O_2 levels are less than 5 mmHg [10,11]. Targeting systems have been designed to measure O_2 concentrations within the clinically relevant range of 0–15 mmHg. The measurement or imaging of O_2 concentrations can be performed three ways: (1) with a ratiometric sensing probe constructed from an O_2-sensitive indicator and an O_2-insensitive dye; (2) with the formation of FRET pairs between donor emission and acceptor absorption bands; and (3) with phosphorescence lifetime imaging. Because hypoxia plays a vital role in cancer

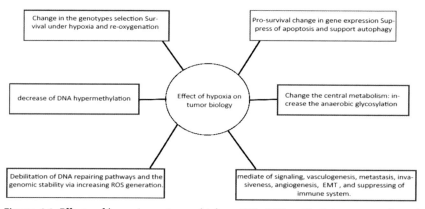

Figure 4.1 Effects of hypoxia on tumor biology [12]. *EMT,* epithelial-to-mesenchymal transition; *ROS,* reactive oxygen species; *TKR,* tyrosine kinase receptor.

progression, detecting and measuring it can effectively detect cancer [12]. One study of hypoxia imaging used a nanoprobe prepared from a poly(N-vinylpyrrolidone) (PVP)-conjugated iridium(III) complex [5]. PVP improved the retention time via the enhanced permeability and retention (EPR) effect and allowed for continuous monitoring of tumor hypoxia. The use of an iridium(III) complex extended phosphorescence emission (PPE) into the near-infrared (NIR) region, improving the penetration depth of the light. The concentration of O_2 in normal tissue is high, causing PPE to be quenched, but in cancerous tissue, hypoxia causes activation of PPE [5].

While the low concentration of O_2 does decrease cancer cell proliferation in the center of a solid tumor, the reduced O_2 levels also provide a suitable habitat for anaerobic bacteria to proliferate in hypoxic tumors. The Luo group engineered anaerobic bacteria, including Bifidobacterium breve and Clostridium difficile, to serve as cargo-carrying (upconversion nanorods) and antibody-directed (Au nanorod delivery) vehicles for imaging and photothermal ablation of tumors (Fig. 4.2) [13]. The in vivo results showed that the antibody-directed strategy had longer retention and more effective imaging and therapy than the cargo-carrying strategy. Wang and coworkers developed a new bimetallic and biphasic Rh-based core—shell nanosystem (Au@Rh-ICG-CM) for addressing tumor hypoxia while achieving high photodynamic therapy (PDT) efficacy. Such porous Au@Rh core—shell nanostructures (NSs) are expected to exhibit catalase-like activity to efficiently catalyze oxygen generation from endogenous hydrogen peroxide in tumors. Coating Au@Rh nanostructures with tumor cell membrane (CM) enables tumor targeting via homologous binding. As a result of the large pores of Rh shells and the trapping ability of CM, the photosensitizer indocyanine green (ICG) is successfully loaded and retained in the cavity of Au@Rh-CM. Au@Rh-ICG-CM shows good biocompatibility, high tumor accumulation, and superior fluorescence and photoacoustic imaging properties. Both in vitro and in vivo results demonstrate that Au@Rh-ICG-CM can effectively convert endogenous hydrogen peroxide into oxygen and then elevate the production of tumor-toxic singlet oxygen to significantly enhance PDT. As noted, the mild photothermal effect of Au@Rh-ICG-CM also improves PDT efficacy. By integrating the superiorities of hypoxia regulation function, tumor accumulation capacity, bimodal imaging, and moderate photothermal effect into a single nanosystem, Au@Rh-ICG-CM can readily serve as a promising nanoplatform for enhanced cancer PDT [14].

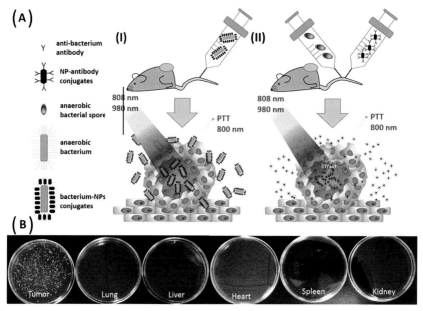

Figure 4.2 *Use of anaerobic bacteria to target tumors.* (A) The scheme shows two approaches involving anaerobic bacteria to deliver functional nanoparticles (NPs): (I) a cargo-carrying method using direct conjugation of NPs to *Bifidobacterium breve* bacteria, and (II) an antibody-directed method involving conjugation of anti-*Clostridium* polyclonal antibodies onto NPs to trigger germination of *Clostridium* spores. (B) Selective growth of *B. breve* in tumor tissues. Tumor-bearing mice were intravenously injected with *B. breve* and sacrificed after 2 days. The tumor tissues and five major organs (lung, spleen, heart, liver, and kidney) were cultured under an anaerobic environment at 37°C. *(Adapted with permission Luo CH, et al. Bacteria-mediated hypoxia-specific delivery of nanoparticles for tumors imaging and therapy. Nano Lett 2016;16(6):3493—99. Copyright 2016, American Chemical Society).*

Fan and coworkers developed multifunctional dendrimer–based nano-sensitizers for dual–mode computed tomography (CT)/magnetic resonance (MR) imaging-guided sensitized radiotherapy (RT) of tumor hypoxias. By virtue of the intelligent dendrimer platform technology, the CT imaging agent of Au NPs and magnetic resonance imaging (MRI) agent of Gd(III)-chelator complexes can be easily incorporated within the dendrimer platform, while the tumor hypoxia-sensitive agent of Nit can be decorated onto the dendrimer surface, thus allowing for effective targeting of hypoxic cancer cells in vitro and tumor hypoxia in vivo. In addition to the good dual–mode CT/MR imaging performance of tumor hypoxia exerted by the designed dendrimer nanohybrids, the attractive nanohybrids can also be used as effective nanosensitizers to reinforce the RT response of tumor hypoxia via enhanced intracellular ROS generation, enhanced DNA damage, and prevention of DNA repair. Such design of dendrimer-based nanohybrids may be applied to precision-imaging-guided sensitized RT of the hypoxic tumors of different types for theranostic applications [15]. Yujing and coworkers proposed a unique, precise breast cancer therapy by silencing hypoxia-correlated protumorigenic gene using a hypoxia-responsive siRNA NP. They revealed that CDC20 expression was upregulated in the tumor tissue of breast cancer patients and was correlated with tumor hypoxia via bioinformatics analyses. In vitro studies also demonstrated a higher level of CDC20 expression in breast cancer cells under hypoxic conditions. As a proof of concept, an NI-modified polypeptide-based hypoxia-responsive nanoparticle (HRNP) was developed for hypoxia-responsive siRNA delivery to target CDC20 specifically. The HRNP/siCDC20 with a polyethylene glycol (PEG)ylated surface and small size showed prolonged blood circulation, high tumor accumulation, highly effective CDC20 silencing, and significant suppression of tumor growth. This combination of dual hypoxia targeting presents a promising and advanced strategy for precision cancer therapy [16]. Zhou and coworkers developed a hypoxia-activatable, cytoplasmic protein-powered fluorescence cascade amplifier for imaging hypoxia associated with inflammatory bowel disease (IBD) in vivo. By virtue of the high capacity of MSNs and the characteristic of the SQ that binds to cytoplasmic proteins accompanied by fluorescence enhancement, this cascade amplifier exhibits high fluorescence enhancement and sensitivity with oxygen levels in the range of 0%–10%. Moreover, the fluorescence imaging results demonstrated that this cascade amplifier could distinguish different levels of cellular hypoxia and monitor variations in hypoxia associated with IBD in the mice model, which could promote our understanding of hypoxia in the development of IBD [17]. Kwon and coworkers designed, developed, and synthesized two novel multifunctional prostate-specific membrane antigen (PSMA) inhibitors containing a PSMA-targeting moiety—one with and one without a hypoxia-sensitive moiety (18F-PEG3-ADIBOT-2NI-GUL and 18F-PEG3-ADIBOT-GUL, respectively; *ADIBOT*: azadibenzocyclooctatriazole, *2NI*: 2-nitroimidazole). Their feasibility as positron emission tomography (PET) tracers for prostate cancer imaging studies was examined. Compounds labeled 18F via the copper-free click reaction were stable in human serum and showed nanomolar binding affinities in in vitro PSMA binding assays. Micro-PET and biodistribution studies indicate that both

18F-labeled inhibitors successfully accumulated in prostate cancer regions. The 18F-PEG3-ADIBOT-2NI-GUL inhibitor showed a twofold higher [tumor]/[total nontarget organ] ratio than 18F-PEG3-ADIBOT-GUL, suggesting that the synergistic effects of the PSMA-targeting GUL moiety and the hypoxia-sensitive 2-NI moiety can increase tumor uptake of novel PET tracers in prostate cancer. These findings suggest that this novel multifunctional PET tracer with an 18F-labeled PSMA inhibitor and a 2-nitroimidazole moiety is a potent candidate to better diagnose prostate cancer via PET imaging studies [18].

Bifunctional therapeutic agents such as Pt(II) porphyrins can be effective agents for the imaging and therapy of cancer under hypoxic conditions. However, Pt(II) porphyrins showed aggregation in aqueous solutions. Addressing this problem, the hydrophilic starburst Pt(II) porphyrins (Pt-1, Pt-2, and Pt-3) with four cationic fluorene oligomeric arms could provide increased water solubility and prevent the aggregation of Pt(II) porphyrins [19]. Among the tested compounds, Pt-3 showed the best results for oxygen-sensing and the highest singlet oxygen quantum yield and was chosen to serve as both a photosensitizer and an oxygen probe for simultaneous PDT and real-time monitoring of cancer hypoxia (Fig. 4.3).

Conventional cancer imaging methods can be invasive and have low specificity and image resolution [20]. The designed hypoxia probe 1 (HyP-1)

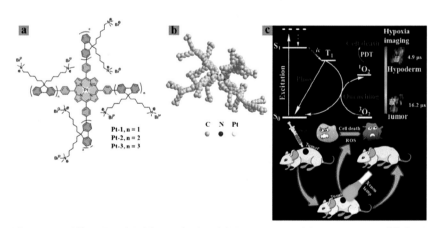

Figure 4.3 *Bifunctional Pt(II) porphyrins.* (A) Pt-1, Pt-2, and Pt-3 structures. (B) Optimized 3-D model of Pt-3 simulated by ChemBio3D. (C) Schematic of Pt-3 used as a bifunctional agent for tumor hypoxia imaging and photodynamic therapy (PDT). *(Adapted with permission Lv Z, et al. Phosphorescent starburst Pt (II) porphyrins as bifunctional therapeutic agents for tumor hypoxia imaging and photodynamic therapy. ACS Appl Mater Interfaces 2018. Copyright 2018, American Chemical Society).*

employing an N–oxide-based trigger could allow facile bioreduction mediated by heme proteins (such as the CYP450 enzyme) in the absence of oxygen and amplify the photoacoustic (PA) signal (Fig. 4.4). The HyP-1 allowed the production of a spectrally distinct signal for PA imaging. In vitro and in vivo results showed that HyP-1 had good selectivity for cancerous tissue in hypoxia conditions and could be used as a multimodality imaging agent [20]. A study described a tunable nanocluster to achieve deeper tumor penetration, a "bomb-like nanoprobe" equipped with active hypoxia-targeting and passive tumor accumulation capability, an initial size of 33 nm, and a long half-life during blood circulation to release small-molecule-based hypoxic microenvironment targeting. CT imaging was assessed in animal models of pancreatic and breast cancer and supported the feasibility of deep hypoxic tumor targeting [21]. The Zhang research team proposed a new multifunctional theranostic agent based on a tumor-targeting photosynthetic biohybrid nanoswimmer system. The new agent was fabricated by integrating biological *Streptomyces platensis* with low-cytotoxicity magnetite NPs, and the result was biocompatible, cost-effective, and reliable for large-scale production. Based on in vitro and in vivo observations, the fabricated microswimmers exhibited trimodal FL/PA/MR imaging and high tumor accumulation potency via magnetic actuation. More importantly, the intrinsic characteristics of photosynthesis and chlorophyll, as well as the engineered tumor–targeting coating, allow the designed MMP-sensitive peptide (MSP) to efficiently produce O_2 to ameliorate tumor hypoxia and enhance RT efficacy while making laser

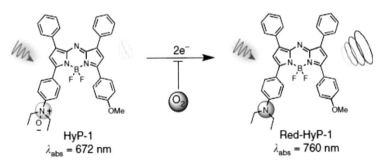

HyP-1
λ_{abs} = 672 nm

Red-HyP-1
λ_{abs} = 760 nm

Figure 4.4 *N-oxide-based hypoxia probe.* Green HyP-1 N-oxide undergoes irreversible two-electron reduction by heme proteins (such as the CYP450 enzyme) in the absence of O_2, which binds competitively to the heme iron. Red-HyP-1 amine generates an enhanced photoacoustic signal (*blue circles*) upon irradiation at 770 nm (*red arrow*). (*Adapted with permission. Knox HJ, et al. A bioreducible N-oxide-based probe for photoacoustic imaging of hypoxia. Nat Commun 2017;8(1):1794. Copyright 2017, Nature*).

irradiation more powerful against highly generated cytotoxic ROS, thus reflecting the strong properties of PDT. Therefore, the biohybrid microswimmers reported herein could be used as a tumor-targeting oxygenator for in situ oxygen generation to regulate the hypoxic tumor microenvironment for FL/PA/MR imaging-guided diagnosis and synergistic RT/PDT treatment of tumors [22].

3.2 pH targeting

Mammalian cells import glucose as the primary source of energy metabolism. There are two possible known metabolic pathways, the Pasteur effect and the Warburg effect. In the Pasteur Effect, glycolysis is inhibited by oxygen, which allows for the glucose metabolite pyruvate to be converted into H_2O, ATP, and CO_2 by oxidative phosphorylation. The Warburg effect occurs under low oxygen conditions and involves aerobic glycolysis that converts glucose into lactic acid. Both these metabolic pathways are essential for maintaining energy and control of pH within the normal range (7.3−7.4) in the extracellular space. In hypoxic cancerous tissue, elevated glucose uptake and increased glycolysis lead to greater lactic acid production. Most glucose is converted into lactate, H^+, and ATP. The produced lactate and H^+ are exported into the extracellular space via the monocarboxylate transporter and the sodium-hydrogen exchanger, respectively, resulting in a reduced pH range (6.2−6.9). The reduced pH of the TME induces tumor progression, enhanced angiogenesis, metastasis, migration, invasion, and mutagenesis, and it inhibits tumor cell apoptosis and antitumor immune response (reviewed in Refs. [23,24]). Within cancerous tissue, extracellular pH (pH_e) varies by the tumor's type, mass, and location (i.e., site) within the body (Table 4.1).

The different pH ranges found in cancerous and normal tissues (Table 4.1) can be used for cancer targeting. The high stability of pH–sensitive NSs in the normal physiological pH range makes them effective targeting strategies for cancerous tissue. When the pH trigger point is reached, the cargo is rapidly released. The following approaches have been developed to achieve this goal: (1) the use of ionizable chemical groups, such as amines, carboxylic acids, and phosphoric acids that can be incorporated in organic (polymers, lipids, and peptides), inorganic (zinc oxide and calcium phosphate), and hybrid nanomaterials; (2) the use of acid-labile chemical linkers such as imine, cis-aconyl, orthoester, and hydrazone, which are covalently attached to the contrast agent and stable when pH is neutral but hydrolyzed or degraded in acidic conditions; (3) carbon dioxide-generating precursors

Table 4.1 pH$_e$ values for various human tumor xenografts [25].

Xenograft cell line	pH range	Xenograft cell line	pH range
Breast cancer		*Lung cancer*	
SE (T60)	6.76—6.84	SE	6.84—6.9
REI	6.78—6.84	KO	6.84—6.97
JE	6.8—6.84	SCHRO	6.68—6.84
GA	6.78—6.84	A 549	6.76—6.84
BR	6.7—6.84	LX-1	6.84—6.9
CH	6.84—6.89	LXFA 289	6.74—6.84
MX-1	6.78—6.9	LXFE 229	6.79—6.84
Miscellaneous		SCLC	6.84—6.89
F8	6.84—6.96	*Sarcoma*	
(neurofibrosarcoma)			
STO (pancreas)	6.72—6.84	BO (osteogenic)	6.75—6.84
LA (endometrium)	6.79—6.84	N4 (malignant fibrous	6.84—6.91
		histiocytoma)	
GE (thyroid)	6.82—6.84	*Gastrointestinal cancer*	
MRI-H-212/B	6.84—6.9	CXF 1103 (colon)	6.84—6.97
(melanoma)			
H-MESO	6.84—6.94	WiDr (colon, adenoma)	6.74—6.84
(mesothelioma)			
Arterial blood	7.36—7.44	SP (stomach)	6.84—7.01

that can react at low pH ($HCO_3^- + H^+ \rightarrow H_2CO_3 \rightarrow H_2O + CO_2\uparrow$) to produce carbon dioxide gas, leading to nanocarrier disintegration and the release of contrast agents [26]; and (4) the use of pH-activatable contrast agents, such as asymmetric cyanine LS662, that are primarily synthetic organic chemical compounds, and as fluorescence dyes, can take on active (on) and inactive (off) conformations. Under normal physiological conditions, these materials have an inactive conformation, but the conformation is changed to the active form as soon as they enter the cancerous tissue. These materials can simultaneously function as both a contrast and a targeting agent [27,28].

Yasuteru et al. [29] developed a tunable and pH-activatable fluorescent probe. They used 2,6-dicarboxyethyl-1,3,5,7-tetramethyl boron-dipyrromethene (BODIPY) as a fluorophore to tune the pH profile of the fluorescent probe and alter the functional group to the aminophenyl BODIPY. The designed probe could be used within a pH range of 2 to 9 for various purposes in in vitro and in vivo studies. Wang et al. [30]

employed a protonatable strategy and prepared micellar NSs as ultra-pH-sensitive (UPS) nanoprobes for extracellular tumor imaging. The designed UPS nanoprobe, with an ultra-pH-sensitive core (poly [ethylene glycol]-b-poly [2-(hexamethylenediamine) ethyl methacrylate] copolymer), had a sharp tunable pH (<0.25) response with a near-infrared fluorescence (NIRF) dye (Cy5.5) as the fluorophore and a targeting agent—Arg-Gly-Asp (RGD)—that bound to the $\alpha v \beta 3$ integrin. At physiological pH (7.4), this fluorescent nanoprobe was self-quenched. At acidic pH (6.9), the UPS nanoprobe showed a sharp and rapid response. The copolymer became protonated, the micelle was disrupted, and the fluorescent dye was activated (Fig. 4.5).

The development of a pH-activatable nanoprobe using PEGylated Mn^{2+}-doped calcium phosphate NPs and poly(ethylene glycol)-b-poly (glutamic acid) (PEG-b-P (Glu)) block copolymers with improved mechanical properties has been reported on previously. The designed pH-sensitive MRI nanoprobes rapidly amplified MR signals under pathological pH conditions. In acidic solid tumors, the designed NPs disintegrated and released Mn^{2+} ions. The relaxivity of Mn^{2+} after binding to the proteins was enhanced, which produced enhanced MRI contrast [31]. The Li research team established a chitosan-based nanocomplex, CE7Q/CQ/S, to deliver the molecular-targeted drug erlotinib (Er), survivin shRNA-expressing plasmid, and photothermal agent heptamethine cyanine dye (Cy7) in one platform for simultaneous NIRF imaging and triple-combination therapy of non-small-cell lung carcinomas (NSCLCs) bearing EGFR mutations. The obtained CE7Q/CQ/S exhibited favorable photothermal effects, good DNA binding ability, and pH/NIR dual-responsive release behaviors. The conjugated Er could mediate specific delivery of Cy7 to EGFR-mutated NSCLC cells to enable targeted NIRF imaging and photothermal therapy (PTT). The in vitro and in vivo results showed that downregulation of Survivin expression and the photothermal effects could act synergistically with Er to induce satisfactory anticancer effects in either Er-sensitive or Er-resistant EGFR-mutated NSCLC cells. By integrating chemo/gene/photothermal therapies into one theranostic nanoplatform, CE7Q/CQ/S could significantly suppress EGFR-mutated NSCLC, indicating its potential use in treating NSCLC [32]. Feng and coworkers introduced a new nanophotosensitizer design in which luminescence upconversion NPs loaded with photosensitizers are self-assembled into a nanoball with the aid of a specific pH-sensitive polymer layer containing overloaded photosensitizers and quenching molecules. This design

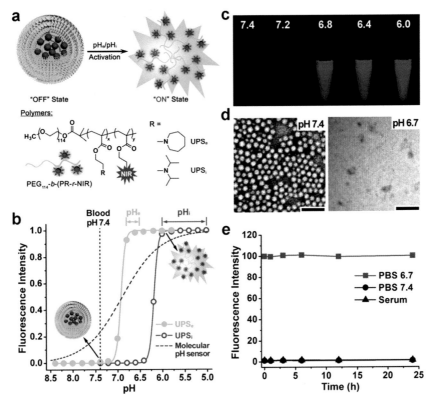

Figure 4.5 *Preparation and characterization of an ultra-pH-sensitive nanoprobe.* (A) Structural composition of two types of nanoprobes, ultra-pH-sensitive extracellular (UPSe) and ultra-pH-sensitive intracellular (UPSi), with pH transitions at 6.9 and 6.2, respectively. The UPSe was specifically designed to be activated in tumor extracellular fluid (pH 6.5−6.8). The UPSi was activated inside acidic endocytic organelles (pH 5.0−6.0). Cy5.5 was used as a near-infrared fluorescence agent in animal studies. (B) Normalized fluorescence intensity as a function of pH for UPSe and UPSi nanoprobes. At high pH (7.4), both probes remained quenched. At pH below their transitions (6.9 and 6.2), micellar dissociation could activate the probes. The *blue dashed line* simulates the pH response of a small molecular pH sensor with a pKa of 6.9 based on the Henderson−Hasselbalch equation. For the UPS nanoprobes, the pH response was extremely sharp. In contrast, small molecular pH sensors require 3 pH units for a comparable signal change. (C) Fluorescent images of UPSe−Cy5.5 nanoprobe solution with various pH buffers. (D) Transmission electron micrographs of UPSe nanoprobes at normal pH and pH 6.7. (E) UPSe nanoprobes remain stable in fresh mouse serum over 24 h at 37°C. *(Reproduced with permission Wang Y, et al. A nanoparticle-based strategy for the imaging of a broad range of tumors by nonlinear amplification of microenvironment signals. Nat Mater 2014;13(2):204. Copyright 2014, Nature).*

makes the therapy "off/on" function possible by only imaging during nanoball circulation ascribed to NIR photon upconversion of the nanoballs and pH-sensitive shell. PDT activation occurs solely after the nanoballs have been taken up by the cancer cells owing to the acidic microenvironment. This design effectively prevents photodamage to the photosensitizers during the enrichment and targeting process of the tumor, as validated in vitro and in vivo, which enables imaging-guided PDT treatment of deep-seated tumors in a much more relaxed and comfortable way for patients. This patient-friendly nanomaterial construction strategy can be extended to other therapies [33]. The Zhao research team developed novel double-charged pH-responsive fluorescent core/shell hydrogel nanoparticles (VANPs—FS) to construct a nanoprobe for accurate targeting and rapid imaging of tumor tissues. First, via an emulsion-free polymerization, hydrogel NPs of uniform size were prepared—the core was Ps, and the shell was poly(VBTAC-co-AA). Then the shell was conjugated with folic acid (FA) through conventional amidation. Next, AIE fluorescence molecule SDSA was attached to hydrogel NPs (VANPs—FA) by tight electrostatic attraction with dramatically increasing fluorescence efficiency. With proper selection (molar ratio of VBTAC to AA is 1:2), VANPs—FS—1:2 show negative charge and disperse stably at high pH (>7), aggregate at pH near electrical neutrality (6.5), and redisperse in water with a positive charge at low pH (<5). At pH = 7.4, VANPs—FS can penetrate specific tumor tissues through the EPR effect, resist the adsorption of opsonic or other anionic serum proteins in the bloodstream, and clear via the mononuclear phagocyte system. In tumor tissue simulation (pH = 6.5), the zeta potential of VANPs—FS is −3.2 mV, near the isoelectric point. VANPs—FS aggregate immediately at a large size of about 683.0 nm, which can prevent NPs from returning to blood vessels and prolong their retention in tumor tissue. The experimental results also demonstrate that pH-responsive VANPs—FS have fluorescence stability against biomolecular effects and changes in pH. The MTT analysis proves that VANPs—FS have good biocompatibility and low cytotoxicity. What is more, the targeting effectiveness of VANPs—FS—1:2 has been confirmed by CLSM. Accompanied by the high affinity of FA with the CM of HeLa cells, the cell uptake of VANPs—FS—1:2 by HeLa cells is substantially greater than 293 T cells. Combining pH-responsive aggregation, resistance to adsorption of biomolecules of biological macromolecules, stable fluorescence, and the active targeting effect of FA, the prepared VANPs—FA as a nanoprobe can selectively target and perform fluorescent imaging of cancer tissues for future cancer diagnosis and therapy [34].

Cancer detection can be improved using dual-activatable imaging probes. Benedict et al. [35] prepared a pH-activatable fluorescence/MRI dual-modality imaging nanoprobe. They co-encapsulated MnO NPs as an MRI contrast agent and fluorescence quencher with coumarin-545T as a fluorophore in hybrid silica nanoshells that were conjugated with FA to target cancer cells (Fig. 4.6). At normal pH, the MnO NPs remained within the nanosystem, and the fluorophore was quenched. In cancerous tissue with low pH, Mn^{2+} was released, which provided strong T_1 contrast enhancement, and in addition, coumarin fluorescence was recovered [35]. Huang and coworkers reported a new method for amplifying imaging contrast in tumors via temporal integration of the imaging signals triggered

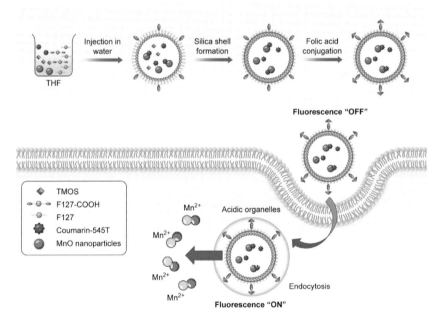

Figure 4.6 Schematic illustration of the preparation and working principle of nanosystems. First, monodisperse MnO NPs (MONPs) were presynthesized. Next, a mixed payload of C545T (as a fluorophore) and MONPs was encapsulated into a carboxylic acid–functionalized silica nanoshell by an interfacial templating scheme. Finally, aminated folic acid was conjugated to the carboxylic acid groups for active targeting of cancer cells. *(Adapted with permission Hsu BYW, et al. pH-activatable MnO-based fluorescence and magnetic resonance bimodal nanoprobe for cancer imaging. Adv Healthc Mater 2016;5(6): 721–729. Copyright 2016, Wiley).*

by tumor acidosis [36]. This method exploits the catastrophic disassembly at the acidic pH of the tumor milieu of pH-sensitive positron-emitting neutral copolymer micelles into polycationic polymers, which are then internalized and retained by the cancer cells. PET imaging of the ^{64}Cu-labeled polymers detected small occult tumors (10−20 mm^3) in the brain, head, neck, and breast of mice at much higher contrast than ^{18}F-fluorodeoxyglucose, ^{11}C-methionine, and pH-insensitive ^{64}Cu-labeled NPs. We also show that the pH-sensitive probes reduce false-positive detection rates in a mouse model of noncancerous lipopolysaccharide-induced inflammation. This macro-molecular strategy for integrating tumor acidosis should enable improved cancer detection, surveillance, and staging.

3.3 Matrix metalloproteinase targeting

MMPs are a family of zinc-containing endopeptidases that play an important role in the degradation of extracellular matrix (ECM) proteins. In normal tissue, MMP expression is regulated by hormones, cytokines, cell−matrix (or cell−cell) interactions, and growth factors. MMPs are present in low quantities, and their activity is regulated by "tissue inhibitors of metalloproteinases" (TIMPs). However, in tumors, the TIMP system becomes dysfunctional, and MMPs (including MMPs 2, 3, 7, and 9) are overexpressed and activated depending on the stage and type of cancer [37−39]. For example, the median concentration of MMP-2 in early-stage ovarian cancer is 0.47 μg/mg, whereas the concentration of MMP-2 in end-stage ovarian cancer is 1.2 μg/mg [40].

While ECM components such as collagen, fibrinogen, and gelatin are natural substrates of MMPs, the large size of these proteins limits their use for targeting applications. MSPs have been used as synthetic MMP sub-strates, as they are composed of the correct amino acid sequence in short linear peptides that are easily incorporated into NSs. The selectivity and specificity of these MSPs depend on the sequence recognized by the spe-cific MMPs [41−43]. Membrane type 1 (MT1)-MMPs are a subfamily expressed on the CM that mediates the pericellular proteolysis and cleavage of cell surface receptors. A study by Kondo et al. [44] used radiolabeled ^{18}F-BODIPY650/665, an MT1-MMP peptide substrate coated with PEG to prevent cell uptake. The MT1-MMP peptide substrate was cleaved by MT1-MMPs, and the PEG moiety was eliminated, allowing accumulation of the probe inside the tumor cells. This ^{18}F-BODIPY650/665 could be used for dual optical imaging and PET. The results showed that MT1-MMPs were active in cancers and could be used as a targeting mo-dality (Fig. 4.7).

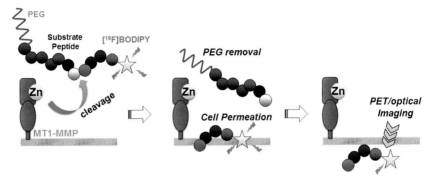

Figure 4.7 *Mechanism of MT1-MMPs used as a targeting agent for cancer imaging.* Because the MT1-MMP substrate peptide is cleaved by membrane type 1 matrix metalloproteinase in tumors, the poly(ethylene glycol) moiety is eliminated, thus allowing the probes to accumulate in tumor cells because the high cellular membrane permeability of ^{18}F-BODIPY can be used for tumor imaging. *(Adapted with permission Kondo N, et al. Development of PEGylated peptide probes conjugated with 18F-labeled BODIPY for PET/optical imaging of MT1-MMP activity. J Contr Release 2015;220:476–483. Copyright 2015, Elsevier).*

Kuo et al. [45] designed an NIRF-sensitive probe for evaluating MMP-3 activity in an ovarian cancer cell line to detect early-stage ovarian cancer. These workers used cyanine dye as a fluorochrome and the amino terminus as a peptide substrate specific for MMP-3. Exposing the MMP-3 sensitive probe to the MMP-3 enzyme significantly increased NIRF emission intensity. More precise targeting of cancer can be achieved when MMPs are integrated with external/internal-responsive agents. A designed dual-stimulus–responsive fluorescent nanoprobe was fabricated from an asymmetric cyanine used as a pH-sensitive fluorescent dye, glycosyl-functionalized gold nanorods, and a specific peptide sequence as a linker and MMP substrate [28]. The inactive form of the nanoprobe existed at pH 7.4 in the presence of a low concentration of MMPs, while the fluorescence was activated in response to acidic pH and higher levels of MMPs as found in the TME (Fig. 4.8).

3.4 Fibronectin targeting

Fibronectin (FN) is a cell-adhesion glycoprotein found in the ECM and various bodily fluids. FN regulates a wide spectrum of cellular and developmental functions, including growth, migration, proliferation, cell

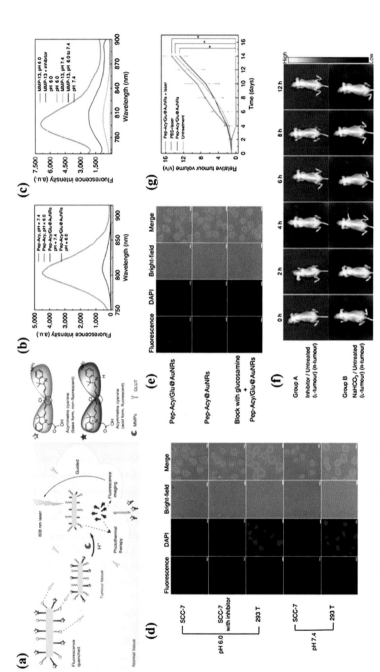

Figure 4.8 *Imaging and photothermal therapy of tumors with a dual-responsive nanoprobe.* (A) Scheme of the nanoprobe as a pH/MMP dual-stimulus responsive pH reversibly activated theranostic platform (Pep-Acy/Glu@AuNRs) for tumor-targeted precision imaging-guided photothermal therapy. (B) Fluorescence spectra of theranostic platform and Pep-Acy. (C) Fluorescence spectra of the theranostic platform by pH for MMP-13. (D) Cell internalization of the theranostic platform in SCC-7, 293 T, and inhibitor pretreated SCC-7 cells. (E) Cell imaging of theranostic platform in SCC-7 cells. (F) Theranostic platform-mediated in vivo fluorescence images in SCC-7 tumor-bearing mice. (G) Comparative tumor volume change in groups of mice. *(Reproduced with permission Zhao X, et al. Dual-stimuli responsive and reversibly activatable theranostic nanoprobe for precision tumor-targeting and fluorescence-guided photothermal therapy. Nat Commun 2017;8: 14998. Copyright 2017, Nature).*

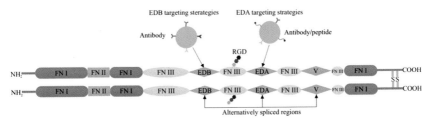

Figure 4.9 *Schematic diagram of fibronectin.* FN is composed of three types of repeats termed FNI (orange), FNII (pink), and FNIII (blue). Three FNIII domains, EDA, EDB, and the V region (light orange), can be alternatively spliced. EDA and EDB domains are markers of angiogenesis, a critical step in tumor progression. EBD targeting strategies consisting of antibody-based delivery (such as L19 and BC-1) and EBA (F9), and peptide-based delivery can be used for therapy, imaging, and vaccination.

adhesion, and wound healing. FN is assembled from monomers consisting of three types of homologous repeat subunits (FNI, FNII, and FNIII domains) with different binding affinities to various ECM proteins [46]. FN contains 12 FNI, 2 FNII, and 15−17 FNIII domains (Fig. 4.9). The two FN subunits are covalently linked via disulfide bonds near their C-terminus. FN has two principal forms: cellular FN, which polymerizes into insoluble fibers in the ECM, and soluble plasma FN. The splicing sites are located in EDA (or EIIIA), EDB (or EIIIB), and IIICS (connecting segment) domains and in regions between domains ^{15}FNIII and ^{14}FNIII. The expression of EDA and EDB domains is extremely restricted in normal human tissue but highly expressed in the ECMs of many cancers.

EDB-FN is absent in adult blood vessels but overexpressed during angiogenesis in normal and neoplastic tissues, making it an attractive marker for angiogenesis [47]. EDA-FN can also act as a marker of normal and tumor vasculature. Oncofetal forms of EDA-FN, EDB-FN, and IIICS-FN have been shown to be overexpressed in various cancers. Changes in FN expression and organization in the ECM contribute to the "premetastatic niche" and may dictate the pattern of metastatic spreading. The deposition of FN in the tumor ECM stimulates the formation of a fibrin−fibronectin complex, which in turn facilitates the proliferation, angiogenesis, and metastasis of cancer [48,49]. During the epithelial-to-mesenchymal transition (EMT), transforming growth factor-beta (TGF-β) increases the expression of FN. The FN abundance can serve as a prognostic biomarker in human cancer. For example, in the case of invasive breast cancer, a significant correlation was found between the FN levels and the pathologic tumor stage, histologic grade, and patient survival rate [50]. Additionally, detection of EDA-FN in urine was shown to be a predictor of survival in bladder cancer patients [51]. Thus, FN is an attractive biomarker for molecular imaging for the early detection of high-risk cancer and for micrometastasis [52]. FN has been used as a target to develop antibody-targeted platforms for accurate and specific delivery of imaging and therapeutic agents to metastatic sites [53,54].

Zhou et al. developed a pentapeptide CREKA-targeted MRI contrast agent CREKA-Tris (Gd-DOTA)$_3$ (Gd-DOTA, 4,7,10-tris (carboxymethyl)-1,4,7,10-tetraazacyclododecane gadolinium) for breast cancer molecular imaging (Fig. 4.10). The CREKA peptide sequence was selectively bound to FN and the fibrin−FN complex. Compared with nontargeted controls, the targeted contrast agents were selective for the ECM of cancerous cells showing good and long-lasting enhancement of tumor contrast. Results showed that the CREKA-targeted imaging construct could act as a noninvasive, high-resolution molecular MRI probe to detect tumor micrometastases (≤ 0.5 mm) [55].

As mentioned above, EDB-FN is an EMT biomarker that can be identified by specific targeting ligands such as the ZD2 peptide sequence (Cys-Thr-Val-Arg-Thr-Ser-Ala-Asp). Han et al. [56] prepared a hydroxylated tri-gadolinium nitride metallofullerene (Gd3N@C80) that acted as a contrast agent, and the ZD2 peptide was used as a targeting ligand in the ZD2-Gd3N@C80 probe, with the ability to detect aggressive tumors using MRI. The MRI data showed the designed probe allowed significantly decreased doses and produced strong signal enhancement in aggressive triple-negative breast cancer (TNBC) in a mouse model [56]. One novel type of potentially clinically translatable molecular-targeted microbubble (MB) preparation included an engineered 10th type III domain of the FN (MB-FN3 VEGFR2) scaffold-ligand to image vascular endothelial growth factor receptor 2 (VEGFR2)-associated neovasculature. The MB-FN3 VEGFR2 was developed for in vivo ultrasound molecular imaging (USMI) of breast cancer neovasculature with specific binding to VEGFR2, which was significantly higher in breast cancer compared with normal breast tissue. The FN3-scaffold could be produced via recombinant technology, with small size, solubility, lack of glycosylation, good stability, and disulfide bonds, leading to the generation of small high-affinity ligands for USMI [57]. Wang and coworkers reported the development of a theranostic collagen-targeted cell-penetrating drug delivery system for the treatment of cardiovascular disease [58]. Caused by the action of MMPs, degraded collagen is a hallmark of unstable atherosclerotic plaques that are highly

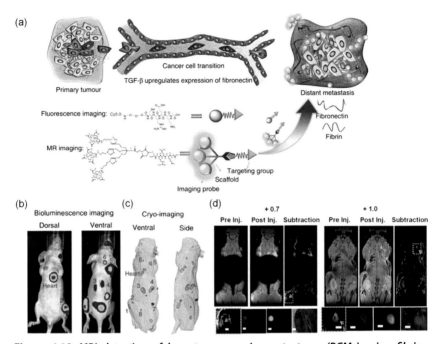

Figure 4.10 *MRI detection of breast cancer micrometastases (BCMs) using fibrin—fibronectin targeted contrast agent, CREKA-Tris (Gd-DOTA) 30.* (A) BC metastasis is accompanied by upregulated fibronectin expression. By targeting overexpressed fibronectin that forms complexes with fibrin, CREKA-Tris (Gd-DOTA), a targeted imaging probe, accumulates at metastasis sites and thus produces tumor contrast enhancement in MRI that was validated by high-resolution fluorescence imaging of CREKA-Cy5 that also accumulated in metastases. (B) MRI images of BCMs contrast-enhanced by targeted imaging probe to show the coronal slices before and after CREKA-Tris (Gd-DOTA) 3 injection, the subtraction images of the preinjection from postinjection images, and the amplified subtraction MRI images of metastatic sites. (C and D). Corresponding GFP cryo-fluorescence images of the micrometastases and CREKA-Cy5.0 images validate MRI detection of micrometastases (tumors are indicated by orange arrow; all scale bars are 1 mm). *(Reproduced with permission Zhou Z, et al. MRI detection of breast cancer micrometastases with a fibronectin-targeting contrast agent. Nat Commun 2015;6:7984. Copyright 2017, Nature).*

susceptible to rupture. Targeting unstable plaques and delivering MMP blockers directly to plaque to inhibit MMP activity is a promising new strategy that requires the benefits and possibilities of nanodelivery approaches. The presented delivery system is designed to (1) target and bind to a cryptic epitope on collagen IV exposed through the degradative action of MMP-2, (2) to image the targeting and cell uptake, and (3) to deliver the

MMP-14 inhibitor naphthofluorescein. In detail, the novel targeting unit is composed of a collagen-homing T-peptide and bound to an MMP-2-cleavable activatable cell-penetrating peptide that, upon cleavage by MMP-2, deposits the MMP-14 blocker drug into cells directly into contact with MMP-14 activating enzymes. To selectively attach both the targeting peptide and a reporting imaging dye, a nanosponge−NP network is modified to present orthogonal aldehyde and thiol functional groups as surface units. The MMP-14 inhibitor naphthofluorescein is loaded into the NP delivery system after postconjugation chemistries and finalizes the synthesis of this novel theranostic delivery system. The ability to evade phagocytosis is confirmed in vitro by using murine RAW cell line, and effective in vitro cell uptake using the MMP-2 producing the HT1080 cell line is demonstrated. In this study, the workers combined a highly specific targeting peptide directed against degraded collagen and a tailorable nanosystem deemed to deliver its potent drug load directly into cells to inhibit the cascade for MMP activation that breaks down collagen structures to rupture plaque, the underlying cause of myocardial infarction and strokes.

Kasten and coworkers [59] designed a peptide probe containing an NIRF dye/quencher pair, a PET radionuclide, and a moiety with a high affinity to MMP-14. This novel substrate-binding peptide allows dual-modality imaging of glioma only after cleavage by MMP-14 to activate the quenched NIRF signal, enhancing probe specificity and imaging contrast. Results showed that novel MMP-14-targeted and activatable peptide probes enabled dual PET and NIRF imaging of glioma in pre-clinical studies. High-NIRF-signal TBRs were observed in the resected brain sections of mice bearing PDX glioma tumors. Correlations between in vivo PET and ex vivo NIRF signals support the concept for dual-modality imaging of glioma with a single, MMP-14-targeted probe scaffold. The colocalization of NIRF signals and MMP-14 expression in the tumors observed by tissue staining confirmed the specific localization of the peptide probes. These results support future preclinical studies designed to test the efficacy of surgical resection of glioma with the MMP-14-targeted probes.

3.5 Apoptosis signature targeting

Therapy methods that function by inhibiting or inducing apoptosis keep growing, imaging systems capable of tracking cell death (apoptosis) will become increasingly crucial. Nowadays, various strategies to monitor apoptosis have been developed on the basis of a wide range of surrogate biomarkers. These include apoptosis signaling molecules such as the caspases and markers downstream in the apoptosis cascade [60]. One of the strategies that can be used for apoptosis monitoring is caspase targeting. Caspases are a family of cysteine proteases–peptidases (Fig. 4.11) that use a cysteine residue as the catalytic nucleophile sharing an exquisite specificity for cleaving target proteins at sites next to aspartic acid residues [61]. The concerted actions of caspases are responsible for apoptosis, a specific form of programmed cell death essential to nematode development and the pathology of many diseases.

The annexin family of membrane-binding proteins can bind to negatively charged phospholipids in the presence of calcium ions. Among the annexin family members, only the V type has an extracellular presence in addition to intracellular localization [62]. During the early phase of programmed cell death, phosphatidylserine (PS) in a lipid bilayer of the CM is flipped from the inner to the outer layer and exposed to the surface.

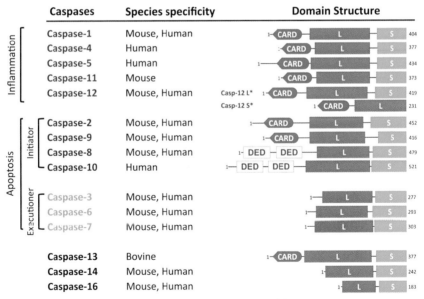

Figure 4.11 Classification of caspase-based structure and function. Apoptotic caspases 2, 8, 9, and 10 are initiators, while caspases 3, 6, and 7 are key executioners. Caspases 1, 4, 5, 11, and 12 are inflammatory. *CARD*, caspase recruitment domain; *DED*, death effector domain; *L*, large subunit; *L**, long form; *S*, small subunit; *S**, short form. *(Reproduced with permission Shalini S, et al. Old, new and emerging functions of caspases. Cell Death and Different 2015;22(4):526. Copyright 2015. Nature).*

Annexin V binds with high affinity to membranes bearing PS. So annexin V is used as a molecular imaging agent to visualize PS-expressing apoptotic cells [63]. Therefore, caspases and annexin V can be used in TME tracking of cell death as a target and targeted agent, respectively.

Deju et al., to evaluate therapeutic efficacy and anticancer drug selection, designed a caspase-sensitive nanoaggregation fluorescent probe (C-SNAF). The designed probe consisted of (1) D-cysteine and 2-cyano-6-hydroxyquinoline (CHQ) moieties linked to an amino luciferin scaffold and (2) an L-DEVD capping sequence and disulfide bond required for two-step activation involving caspase-3/7-mediated cleavage and intracellular thiol-mediated reduction. The results of in vitro and in vivo studies showed that in tumor tissue unresponsive to therapy, procaspase-3/7 dominates and cannot release the L-DEVD (capping peptide) from C-SNAF, resulting in rapid clearance of the probe. However, therapy-responsive tissue exhibited increased CM permeability and extensive activation of caspase-3/7 on the progression to cell death, and the intensity of fluorescence was increased [64]. In another study, Zhang and coworkers designed a real-time apoptosis imaging AuHNRs-DTPP nanoplatform. For this purpose, they used chimeric peptide (DTPP) on the surface of Au hollow nanorod for NIR-II tumor photothermal therapy, real-time apoptosis imaging, and supplementary PDT. Under laser irradiation, AuHNRs-DTPP exhibits high photothermal conversion efficiency. The results showed that photosensitizer in DTPP was quenched after being loaded onto the surface, but upon AuHNRs-DTPP nanocarrier exposure to caspase-3, the photosensitizer could be released and activated with enhanced fluorescence for apoptosis imaging in vivo and in PDT [65].

The Lu research team designed a novel annexin V labeled with NOTA-maleimide aluminum [^{18}F] fluoride complexation and evaluated it as a novel apoptosis targeting agent in vitro and in vivo. The study results showed that the rate of the tracer bound to erythrocytes with exposed PS was 89.36%. So the designed probe has excellent specificity in apoptotic cell targeting of ^{18}F—AlF-NOTA-MAL-Cys-Annexin V, making it suitable for further investigation in clinical apoptosis imaging [66]. Moreover, annexin V forms labeled with various types of radionuclides are useful as radiotracers for in vivo tracing of apoptosis in SPECT and PET imaging agents. Annexin V and Annexin V derivatives were radiolabeled with ^{111}In, ^{123}I, and ^{125}I for SPECT imaging of apoptosis [67—69].

3.6 Vasculature targeting

A tumor cannot grow without a blood supply. Tumors take over existing blood vessels and stimulate the angiogenesis of new vessels to secure their supply of blood [70]. The angiogenic switch is an important early event in tumor progression: neovascularization beginning in premalignant lesions. Under normal conditions, angiogenesis occurs in inflammatory conditions, in tissue regeneration, and with cancer. Tumor angiogenesis is initiated by local hypoxia and then continues with the expression of other targetable factors, such as VEGFs, VEGFRs, platelet-derived growth factors (PDGFs), angiopoietins, ephrins (EPH receptors), integrins (especially $\alpha v\beta 3$ and $\alpha v\beta 5$), and endoglin (CD105), to recruit supporting cells. In addition to the maturing of new vessels, the endothelial tubes acquire supporting cells such as PERICYTES, smooth muscle cells, and ECM. Tumor vessels are leaky and tortuous, their diameter is irregular, and their walls are thin. Deficiency of pericyte, or pericyte function, could be responsible for these morphological features in tumor vasculature. Therefore, angiogenesis and the factors it involves are suitable candidates for targeted cancer imaging and the evaluation of responses to therapy [71,72]. The de Bruijn team obtained insights with unprecedented detail for targeted PDT, which are significant at present because EGFR-targeted PDT using antibodies as carriers is currently being tested in phase 1/2 clinical trials. NB—PS conjugates 7D12-PS and 7D12-9G8-PS, targeting EGFR, showed significant tumor localization in vivo, which was higher for 7D12-9G8-PS. The significantly higher tumor colocalization combined with prolonged fluorescence intensity time after administration makes 7D12-9G8-PS a good candidate for fluorescence image-guided surgery. Illumination 1 h after the administration of either conjugate resulted in significant tumor necrosis. Despite the difference in fluorescence intensity, no significant difference was observed in overall acute tissue response; both EGFR-targeted NB—PS-mediated PDT treatments led to similar tumor necrosis 2 days post treatment with no notable differences in vascular response [73]. Paiva et al. prepared surface-modified micelles with peptide GE11 for targeting the EGFR [74]. In vitro fluorescence studies demonstrated significantly higher internalization of GE11 micelles into EGFR-expressing HCT116 colon cancer cells versus EGFR-negative SW620 cells. Azo coupling chemistry of tyrosine residues in the peptide backbone with aryl diazonium salts was used to label the micelles with radionuclide ^{64}Cu for PET imaging. In vivo analysis of ^{64}Cu-labeled micelles showed prolonged blood circulation and predominant hepatobiliary clearance. The biodistribution profile of

EGFR-targeting GE11 micelles was compared with nontargeting HW12 micelles in HCT116 tumor-bearing mice. PET revealed increasing tumor-to-muscle ratios for both micelles over 48 h. Accumulation of GE11-containing micelles in HCT116 tumors was higher compared with HW12-decorated micelles. Our data suggest that the efficacy of image-guided therapies with micellar NPs could be enhanced by active targeting, as demonstrated with cancer biomarker EGFR.

Hao et al. designed radiolabeled a NOTA-GO-TRC105 nanoprobe for neovasculature targeting in tumor masses. For this purpose, they used TRC105 as an antibody to CD105 targeting. Pharmacokinetics and tumor-targeting efficacy of the graphene oxide (GO) conjugates were investigated with serial noninvasive PET imaging and biodistribution studies, which were validated by in vitro, in vivo, and ex vivo experiments. The study results introduced CD105 as a promising vascular target for cancer [75]. Gao and coworkers designed, synthesized, and surface engineered a tumor neovasculature targeting "nanobomb." This "nanobomb" was rationally fabricated via encapsulation of vinyl azide (VA) into c(RGDfE) peptide-functionalized, hollow copper sulfide (HCuS) NPs. The resulting RGD@HCuS(VA) was selectively internalized into integrin $\alpha v\beta 3$-expressing tumor vasculature ECs and dramatically increased the photoacoustic signals from the tumor neovasculature, achieving a maximum signal-to-noise. The results show that the designed probe produces high-resolution photoacoustic angiography combined with excellent biodegradability and facilitates the precise destruction of tumor neovasculature by RGD@HCuS(VA) without damaging normal tissues. So this nanobomb has great potential for clinical translation to treat cancer patients with NIR laser-accessible orthotopic tumors and therapy result tracing [76]. For monitoring responses to anticancer therapy, Korpanty and coworkers designed targeted MBs that can be used to effectively monitor response to different therapeutic regimens in models of pancreatic cancer. The result of this study showed that contrast signal using MBs targeted to endoglin (CD105), VEGFR2, or the VEGF-VEGFR complex correlates with immunohistochemically assessed vascular expression of these markers and with tumor microvessel density. Thus, they proposed that ultrasonic imaging using targeted MBs could be used to detection of tumor angiogenesis and assessment of vascular markers expression in response to therapy [77]. Cetuximab–dye conjugates have shown great potential for image-guided surgery of EGFR-positive cancers in clinical trials. However, their long circulation half-life and prolonged generation of high background signals require the injection of antibody conjugates several days prior to imaging, which limits the clinical applications. For this purpose, Kim et al. developed a cetuximab—ATTO655 conjugate (i.e., Q-Cetuximab) for fast and real-time fluorescence imaging of EGFR-positive lung cancers. The fluorescence intensity of Q-Cetuximab was quenched to just 6.9% of that of the unconjugated dye when only 2.14 ATTO655 dyes were conjugated to cetuximab. In vitro real-time cell imaging showed that EGFR-positive A549 cells emitted strong fluorescence at 10 min after Q-Cetuximab treatment in the absence of the washing step, implying target-specific activation of quenched Q-Cetuximab fluorescence upon binding with EGFR-positive cancer cells. When mice with orthotropic A549 tumors received intravenous injection of Q-Cetuximab, scattered microsized tumors in the lungs could be clearly identified from NIRF imaging with a tumor-to-background ratio of 4.28 at 8 h post injection. For comparison, the cetuximab—Alexa647 conjugate (i.e., ON-Cetuximab), which does not show fluorescence quenching, was synthesized as an always-on type of probe. The ON-Cetuximab-treated mice expressed strong fluorescence throughout their body at 8 h post injection; therefore, lung tumor sites could not be discriminated using fluorescence imaging. These results confirm the benefits of Q-Cetuximab for image-guided precision surgery of EGFR-positive lung cancers [78].

4. Endothelial cell surface factors targeting

4.1 Vascular endothelial growth factor targeting

VEGF (known as a tumor vasculature marker) conducts EC proliferation, survival, migration, invasion, vascular permeability, chemotaxis of bone marrow-derived progenitor cells, and vasodilation [79]. The VEGF family consists of VEGFA (commonly referred to as VEGF), VEGFB, VEGFC, VEGFD, and placenta growth factor, glycoproteins [80,81]. The VEGF ligand has three types of VEGF receptors: VEGFR1 (for VEGF), VEGFR2 (a positive regulator of angiogenesis), and VEGFR3 (for VEGFC and VEGFD) [82,83]. VEGFRs are expressed in normal tissues in a controlled manner. However, in tumor tissue, the expression is strongly upregulated on the surface of ECs [84]. This localization can be used in the targeting and bioimaging of cancers. Gerber et al. [85] used ^{89}Zr and ^{111}In radiolabeled bevacizumab (a monoclonal antibody (mAb), which binds to all the isoforms of VEGFA ligands). The results of micro-CT and micro-PET imaging showed the imaging probe had a significant tumor uptake in the ovarian xenograft tumor model, compared with nonspecific ^{89}Zr-IgG and ^{111}In-IgG as control groups. ^{89}Zr-bevacizumab not only allowed imaging for up to 168 h but also enabled quantitative measurement of the tumor uptake [86]. In another study, Anton et al. [87] used IR Dye 800CW as a fluorescent dye and ^{89}Zr as a radiolabel, both bound to the bevacizumab antibody. They evaluated the tumor uptake and the optimal time for the imaging to achieve the best contrast. Both in vivo fluorescence and PET imaging showed that the fluorescent-labeled VEGF antibody could mediate the highly specific and sensitive detection of tumors.

Another example was mAb VEGF-targeted bovine serum albumin-coated magnetic NPs (MNP@BSA), which have been used for targeting VEGFR in brain cancer using MRI [88]. The results indicated that MNP@BSA was effective in MRI visualization of intracranial gliomas and could be used as a targeted contrast agent. Additionally, the level of VEGFR expression depended on the type of cancer. For example, the ^{124}I-HuMV833 imaging probe was tested in ovarian and colon cancers. PET imaging results showed that uptake of the targeted imaging probes in ovarian tumors was greater than the uptake in colon cancer [89]. Wang et al. [90] developed a small-sized, bispecific fusion protein that can target both EGFR and VEGF pathways simultaneously in pancreatic ductal adenocarcinoma with high efficiency. The bispecific fusion protein Bi50 with a molecular weight of $\sim$50 kDa was constructed by genetic fusion of the two respective binding domains. The synthesized fusion protein Bi50 showed very good bispecific targeting for VEGF and EGFR simultaneously in vitro and in vivo. Additionally, the fusion protein Bi50 showed increased intratumoral permeability and enrichment characteristics in orthotopic Bxpc3 pancreatic tumors. Moreover, the fusion protein Bi50 not only largely targeted tumor vasculature-rich areas (overexpression of VEGF) but also largely bonded the tumor parenchymal cells (EGFR overexpression), thus achieving a "multilevel" targeting effect. Based on such enhanced targeting effect of Bi50, NIRF imaging-guided delineation of surgical margins during resection was successfully achieved in an orthotopic pancreatic cancer model.

4.2 Integrin targeting

Integrins are a family of transmembrane glycoprotein cell surface receptors that facilitate the bonding of the cell to the ECM and immunoglobulins. These receptors contain 24 heterodimers on the cell surface and are formed from 18 α-subunits and 8 β-subunits. In the TME, integrins encourage tumor progression in several different ways, including tumor cell

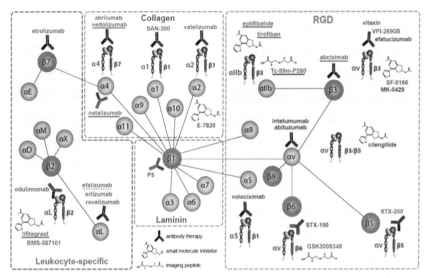

Figure 4.12 *Classification of integrin-based targeting by type of receptors and compounds that target integrins. (Adapted with permission Raab-Westphal S, Marshall JF, Goodman SL. Integrins as therapeutic targets: successes and cancers. Cancers 2017;9(9):110. Copyright 2017, Cancer).*

proliferation, survival, and invasion. Integrins are expressed on fibroblasts, marrow-derived cells, platelets, vascular endothelium, and perivascular cells and facilitate cancer progression [91]. The classification of integrins is dependent on the type of receptors present. Experimental therapeutic compounds involving integrins have been reviewed in this article [92] (Fig. 4.12). Under normal conditions, integrins mediate epithelial cell adhesion to the basement membrane and are usually expressed only at low levels in adult epithelia. However, in epithelial cells originating from solid tumors, integrin expression is altered [93]. The profiles of overexpressed integrins and phenotypes in some human tumors are summarized in Table 4.2.

Different molecular ligands, such as the RGD and Leu-Asp-Val motifs, can be used to target integrin receptors or subunits. The eight families of integrins, which play an important role in cancer progression, can all be targeted with the RGD tripeptide motif [109,110]. The selective accumulation of ^{125}I-RGD-CR780-PEG5K NPs detected by SPECT, CT, photoacoustic, and fluorescence imaging showed that NPs were effective imaging probes and accumulated on αvβ3 integrins expressed in glioblastoma. Furthermore, the data collected from PAI showed that the probe

Table 4.2 Overexpressed integrins and associated phenotypes in some human tumors.

Tumor type	Integrins expressed	Major associated consequence	References
LCBM[a]	αvβ6 and αvβ3	High expression in endothelial cells, low expression in tumor cells	[94]
Prostate	αvβ3 and αvβ5	High expression in peri-tumoral tissue depending on differentiation	[95,96]
Breast	α6β4 and αvβ3	Correlated with increased tumor size and grade and decreased survival (α6β4). Increased bone metastasis (αvβ3)	[97,98]
Pancreatic	αvβ3	Lymph node metastasis	[99]
Glioblastoma	αvβ3 and αvβ5	Both expressed at the tumor −normal tissue margin with a possible role in invasion	[100]
Ovarian	α4β1 and αvβ3	Increased peritoneal metastasis (α4β1) and tumor proliferation (αvβ3)	[101,102]
Cervical	αvβ3 and αvβ6	Decreased patient survival	[103]
NSCLC[b]	α5β1	Decreased survival in patients with lymph node-negative tumors	[104]
Melanoma	αvβ3 and α5β1	Vertical growth phase and lymph node metastasis	[105,106]
Liver	αvβ6	Differentiates cholangiocarcinoma from hepatocellular carcinoma	[107]
Colon	αvβ6	Reduced patient survival	[108]

[a] Lung cancer brain metastases.
[b] Non-small-cell lung carcinoma.

selectively targeted angiogenic tumor vessels [111]. In another study, conjugated quantum dot (QD)-cyclic RGD peptides (D-phenylalanine-lysine) (cRGDfks) were used for targeting the αvβ3 integrin. Fluorescence imaging showed that cRGDfk-QDs had a highly selective uptake in tumor cells and tissues [112]. Recent developments in NIR dyes and imaging modalities enable tumor fluorescent images in preclinical and clinical settings. However, NIR dyes have several drawbacks, and therefore, there is an unmet diagnostic need for NIR dye encapsulation in appropriate pharmaceutical nanocarriers with targeting abilities for the purpose of achieving effective diagnosis and image-guided surgeries. Because integrin receptors are established diagnostic targets, the RGD peptides, recognizing the αVβ3 integrin, have been extensively investigated for radiology and bioimaging of tumors. However, the Lys(Arg)-Thr-Ser [K(R)TS] cyclic peptides, selective for collagen receptors α1β1/α2β1 integrins, which are overexpressed in many tumors, were not yet investigated and therefore were used here for tumor bioimaging with a unique α2β1-integrin-targeted nanocarrier, encapsulating the indocyanine green NIR dye. To solve this problem, Fluksman et al. synthesized three kinds of peptides: two cyclic Arg-Thr-Ser (RTS) peptides functional only in the cyclic conformation and a linear peptide lacking the cyclic cysteine constrained RTS loop. We used them for the preparation of integrin-targeted self-assembled nanocarriers (ITNCs), referred to as OF5 and OF27, and a nontargeted control nano-carrier referred to as OF70. Their selective association was demonstrated with α2β1 integrin expressing cell cultures and three-dimensional tumor spheroids and by competition with an α2β1 selective disinterring. Cytotoxicity experiments in vitro demonstrated the safety of the ITNCs. The targeting potential and the biodistribution of the ITNCs, applied intravenously in A431 tumor-bearing nude mice, were evaluated in vivo using NIR bioimaging. Time-dependent biodistribution indicated that the ITNC OF27 showed higher fluorescent signals in main tissues, with no cytotoxic effects to major organs, and presented higher accumulation in tumors. Cumulatively, these results highlight the potential of the ITNC OF27 as an optical and innovative pharmaceutical bioimaging system, suitable for integrin α2β1 receptor in vivo tumor targeting and visualization in the NIR region [113].

Zhao et al. reported a novel second near-infrared (NIR-II) fluorescent probe QT-RGD constructed with an NIR-II emissive organic fluorophore and two cRGD peptides that can specifically bind to the tumor-associated αvβ3 integrin for accurate tumor diagnosis and targeting therapy. The isotopic ^{125}I-labeled probe exhibited great tumor-targeting ability and emitted intensive NIR-II/PA/SPECT signals, which allows specific and sensitive multimodal visualization of tumors in vivo. More notably, this probe could also be applied for effective imaging-guided PTT of tumors in mouse models owing to its prominent photothermal conversion efficiency and excellent photothermal stability. They thus envision that our work, which unveils a combination of NIR-II/PA/SPECT imaging and PTT, would offer a valuable means of improving tumor diagnostic accuracy as well as therapeutic efficacy [114]. Li and coworkers developed a ^{68}Ga-radiolabeled peptide tracer targeting the α3 unit of VLA-3 and evaluated its potential application in PET imaging of pancreatic cancer.

NOTA-CK11 was prepared by solid-phase synthesis and successfully radiolabeled with 68Ga with greater than 99% radiochemical purity and specific activity of 37 ± 5 MBq/nmol (n = 5). The expression level of integrin α_3 in three human pancreatic cancer cells was evaluated with the order of SW1990, BXPC-3, and PANC-1 from high to low, while the expression level of integrin β_1 was relatively close. When SW1990 cells with the highest expression level of VLA-3 were stained with FITC— CK11, strong fluorescence was observed by flow cytometry and under a laser confocal microscope. However, no significant fluorescence was observed in the blocking group when treated with excessive CK11. ^{68}Ga-NOTA-CK11 showed significant radioactivity accumulation in SW1990 cells and was blocked by CK11 successfully. Subsequent small-animal PET imaging and biodistribution studies in mice bearing SW1990 xenografts confirmed its high tumor uptake with a good tumor-to-blood ratio and tumor-to-muscle ratio (2.45 ± 0.31 and 3.65 ± 0.33, respectively) at 1 h post injection of the probe. In summary, they successfully developed a peptide-based imaging agent, ^{68}Ga-NOTA-CK11, that showed a strong binding affinity with VLA-3 and good target specificity for SW1990 cells and xenografted pancreatic tumor, rending it a promising radiotracer for PET imaging of VLA-3 expression in pancreatic cancer [115]. Kim et al. proposed a zwitterionic NIR fluorophore—tryptophan (Trp) conjugate with a cleavable linker as a minimal-sized versatile platform (MP) for the preparation of peptide ligand-based off—on types of molecular probes. The zwitterionic NIR fluorophore in MP undergoes fluorescence quenching via a photo-induced electron transfer mechanism when in close proximity to tryptophan, and nonspecific binding with serum proteins is minimized as the fluorophore becomes more zwitterionic. The linker can be cleaved inside cancer cells in response to tumor-associated stimuli. As a proof-of-concept experiment, ATTO655 was covalently linked with Trp via a di-arginine linker to form an MP. A cyclic peptide consisting of Arg-Gly-Asp-d-Phe-Lys (cRGD) was used as a cancer-targeting ligand and was conjugated to the MP to form cRGD-MP. The NIRF of cRGD-MP could be selectively turned on inside the target cancer cells, thereby enabling specific fluorescence imaging of integrin $\alpha v \beta 3$-overexpressing cancer cells in vitro and in vivo [116].

More precise targeting of integrins could be achieved by designing ^{18}F-FB-PEG$_3$-GLU-RGD-BBN for dual targeting of the gastrin-releasing peptide receptor (GRPR) and integrin $\alpha v \beta 3$. The results showed that this imaging probe had high tumor accumulation with a favorable pharmacokinetic profile [117]. The binding of FN (a natural ligand) to α5β1 integrin requires the involvement of two small peptide sequences: PHSRN (Pro-His-Ser-Arg-Asn, synergistic binding site) and RGD (primary binding site). Zhao et al. [118] functionalized an α5β1-specific small peptide sequence that acted as an FN mimetic and PR-b (KSSPHSRN (SG)₅ RGDSP), which was modified with β-alanine residues, conjugated to p–SCN–Bn-NOTA and radiolabeled with ^{18}F as a PET imaging probe. Both the imaging and biodistribution results suggested there was higher uptake of the designed probe in α5β1-positive tumors, compared with α5β1-negative tumors; and higher α5β1-positive tumor uptake of the designed probe compared with the control probe. There was no significant difference between the designed and control probes in the uptake into the contralateral muscle.

4.3 Vascular cell-adhesion molecule-1 targeting

VCAM-1 (CD106) was expressed on human CD34 hematological precursor cells and mediated their homing in the bone marrow stroma [119]. VCAM-1 was also expressed on the lateral and luminal side of ECs, and mediated extravasation of leukocytes in inflammatory conditions [120]. Integrins have binding patterns for VCAM-1. Between them, α4β1 is most investigated [121–123]. VCAM-1 has two splice variations in humans, consisting of seven and six Ig-like domains (7 and 6d) [124]. In comparison with VCAM-1 (7d), VCAM-1 (6d) binds to VLA-4 with higher affinity in soluble conditions. In mediating cell separation and adhesion, VCAM-1 (7d) is better and more effective [125].

Under an inflammatory response, VCAM-1 is overexpressed and mediated by ROS, toll-like receptors, agonists, shear stress, cytokines, high glucose concentrations, and oxidized low-density lipoprotein (oxLDL). Tumor tissue expression of VCAM-1 is variable. For example, in ECs and angiogenic vessels, VCAM-1 expression is upregulated and decreased, respectively. Additionally, on the tumor cell surface, VCAM-1 expression is aberrant, while its expression in the lymphatic ECs is constitutive [126]. Though the expression of VCAM-1 in some types of cancer is not certain, circulate cancer cells with adequate VCAM-1 expression levels are promising candidates for targeted cancer imaging and therapy [127]. Micro-PET/CT results from one study showed that ^{68}Ga-NOTA-VCAM-1$_{ScFV}$ has a higher uptake in B16F10 cell line than A375m cells when used as a linear imaging nanoprobe [128]. After confirming the evaluated result, they used LY2409881 as an IKKβ inhibitor (that can induce apoptosis of

VCAM-1-positive cells) and DMSO as a control group. In the control group, uptake of the probe as a tracer consistently remained at the same level. In the treated group, however, uptake of the tracer in the first week decreased and then slowly recovered until it reached the initial level. This study shows that VCAM-1 can be used as a targeting receptor for specific and selective cancer targeting.

As mentioned, VCAM-1 is a transmembrane glycoprotein closely related to tumorigenicity as well as tumor metastasis. It is also a well-known candidate for detecting tumors. LY2409881, an IKKβ inhibitor, could induce apoptosis of VCAM-1-positive cells. For this, Zhang et al. proposed a novel tracer to evaluate the feasibility of detecting VCAM-1 expression and monitoring the LY2409881 tumor curative effect. The tracer was prepared by conjugating the single-chain variable fragment (scFv) of VCAM-1 and NOTA−NHS−ester and then labeled with ^{68}Ga. ^{68}Ga-NOTA-VCAM-1scFv was successfully prepared with high radiochemical yield. VCAM-1 overexpression and underexpression of the B16F10 and A375m melanoma cell lines were used in this study. The results of micro-PET/CT imaging in small animals indicated that the uptake of 68Ga-NOTA-VCAM-1scFv in B16F10 tumor was much higher than that of A375m, and this was confirmed by biodistribution and autoradiography results. LY2409881 inhibits the growth of B16F10 melanoma in vivo by inducing dose- and time-dependent growth inhibition and apoptosis of the cells. The LY2409881 treated group and DMSO control group were established and imaged by micro-PET/CT. In the LY2409881 group, uptake of the tracer in the tumor decreased after the first week and then gradually recovered to the initial level. In DMSO control, the uptake of the tracer remained at the same level during the whole time. The results suggested that LY2409881 inhibits the expression of VCAM-1 and suppresses tumor growth. ^{68}Ga-NOTA-VCAM-1scFv, an easily synthesized probe, has a potential clinical application in the visual monitoring of IKKβ inhibitor intervention on VCAM-1-positive tumors [128]. VCAM-1 induces the inflammation effect in ECs. In one study, Patel et al. [129] used radiolabeled iron oxide NPs conjugated with anti-VCAM-1 antibodies for evaluating the inflammatory TNF-α (tumor necrosis factor-alpha) marker in the rat model of status epilepticus. Imaging results have shown that the contrast agent rapidly and effectively localized binding to the vasculature of the inflamed brain tissue. In addition, the pattern of hypointensity detecting with the MRI was in unanimity with the distribution of the contrast agent as a resolute with phosphor-imaging and SPECT.

Neurite outgrowth is the critical step of nervous development. Molecular probes against neurites are essential for evaluation of the nervous system development, compound neurotoxicity, and drug efficacy on nerve regeneration. To obtain a neurite probe, Wang and coworkers developed a neurite-SELEX strategy and generated a DNA aptamer, yly12, which strongly binds neurites. The molecular target of yly12 was identified to be neural cell-adhesion molecule L1 (L1CAM), a surface antigen expressed in the normal nervous system and various cancers. Here, yly12 was successfully applied to image the three-dimensional network of neurites between live cells, as well as the neurite fibers on normal brain tissue section. This aptamer was also found to have an inhibitory effect on neurite outgrowth between cells. Given the advantages of aptamers, yly12 holds great potential as a molecular tool in the field of neuroscientific research. The high efficiency of neurite-SELEX suggests that SELEX against a subcellular structure instead of the whole cells is more effective in obtaining the desired aptamers [130]. Uddin et al. [131] used the VCAM-1 targeted antisense hairpin and DNA-functionalized gold NP (AS-VCAM-1 hAuNP) for real-time detection and imaging of VCAM-1 expression in retinal ECs. An increase in VCAM-1 mRNA levels caused fluorescence enhancement that was clearly visualized and increased the signal/noise rate. VCAM-1 as a biomarker was also overexpressed in the early development of cancer micrometastases. In the iron oxide imaging probe (VCAM-1-MPIO), microparticles were developed and administered in xenograft models of brain micrometastasis for lung adenocarcinoma, melanoma, and human breast carcinoma as cancer models. The expression of VCAM-1 is more than that of the metastases and was upregulated and independent from the primary tumor type. In addition, MRI imaging results show that VCAM-1R targeting is an excellent strategy for detecting brain cancer micrometastases from three primary cancer types [132]. VCAM-1 is expressed preferentially on the mesothelium of ovarian cancer peritoneal with the potential to act as a marker in cancer metastasis, monitoring, and staging. Scalici et al. [133], designed an SPECT/CT imaging probe using VCAM-1 targeted peptide (tVCAM-4 ([(VHPKQHRGGSPEG5K) 4K] 2-KK (DOTA)-βA-NH$_2$)) and ^{111}In as a radiolabeled agent. In vivo imaging results showed a correlation between VCAM-1 expression levels and tumor stages. Clinically relevant imaging probes identified VCAM-1 expression level as an indicator of ovarian cancer peritoneal metastasis and as a therapeutic response to platinum-based agents.

5. Cancer cell surface marker targeting

5.1 Transferrin receptor targeting

The transferrin receptor (TfR) is a homodimer (180 kDa) type II transmembrane glycoprotein that is integrated into the CM and plays an important role in iron uptake and homeostasis and regulates cell growth via interaction with the iron-transporting protein transferrin [134]. Transferrin (Tf) is produced by the liver and transports iron ions in the body. Depending on the tumor cell iron requirements, higher expression of TfR has been shown in many malignancies compared with normally dividing cells (by up to 100-fold) [135]. Thus, using Tf itself or anti-TfR antibodies could be employed to design different targeted theranostic agents for cancer cells. Biocompatible gadolinium biomineralized transferrin NPs (Gd@Tf NPs) were used to enhance T_1 signal amplification for MRI by increasing the tumor-targeting ability [136]. Interestingly, the T_1 relaxivity of Gd@Tf NPs was much higher than that of Magnevist (a commercial MRI contrast agent), which were measured to be $17.42 \text{ mM}^{-1}\text{s}^{-1}$ and $3-5 \text{ mM}^{-1}\text{s}^{-1}$, respectively. This result could be due to the augmentation effect of protein on the relaxivity of Gd ions. Furthermore, compared with nontargeted NPs, Gd@Tf NPs enhanced the amplification of the T_1 MR signal and showed better tumor localization in vivo. Gd@Tf NPs were excreted from the body via the hepatobiliary system.

In another study, Wang et al. [137] developed self-assembled transferrin-IR780 NPs (Tf-IR780 NPs) for targeted imaging and phototherapy in colon cancer cells (CT26) and normal fibroblasts (L929). As expected, CT26 showed a significantly stronger red fluorescence in the cytoplasm compared with L929, indicating the targeting ability of Tf toward overexpressed TfR on the surface of CT26 cells. The in vivo biodistribution profile of Tf-IR780 NPs in CT26-bearing mice demonstrated a strong signal in the tumor area at 12 h post injection and reached its maximum value after 48 h (Fig. 4.13). The ex vivo imaging results revealed the accumulation of Tf-IR780 in tumor sites was much higher than other organs at 24 h post injection.

To achieve liver tumor imaging with bifunctional nanoprobes, Qi et al. [138] encapsulated SPIONs into PEG-poly(ε-caprolactone) (PEG-b-PCL) polymeric micelles that were decorated with Tf and NIRF dye Cy5.5 to produce nanosized SPIO@PEG-b-PCL-Tf/Cy5.5 (SPPTC). No expression of TfR mRNA in the HL7702 normal cell line was observed, while there was good expression in the HePG2 carcinoma cells. The in vivo fluorescence imaging showed strong fluorescence at the tumor site 8 h post

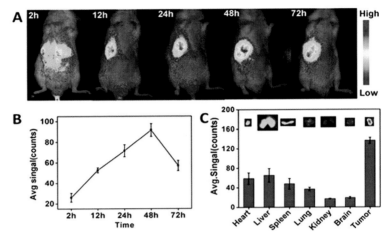

Figure 4.13 *In vivo fluorescence imaging of Tf-IR780 NPs in tumor-bearing mice.* (A) In vivo NIR imaging and (B) NIR intensity values of the mice bearing CT26 tumor injected with Tf-IR780 NPs (0.3 mg/kg, IR780) at 2, 12, 24, 48, and 72 h post injection, respectively; (C) Ex vivo imaging and NIR intensities of Tf-IR780 NPs in heart, liver, spleen, lung, kidney, brain and tumor of the mice bearing CT26 tumor at 24 h post injection. *(Adapted with permission Wang K, et al. Self-assembled IR780-loaded transferrin nanoparticles as an imaging, targeting and PDT/PTT agent for cancer therapy. Sci Rep 2016;6:27421. Copyright 2016, Nature).*

injection of SPPTC, while there was no detectable signal in mice injected with nontargeted NPs (SPPCs). To evaluate SPPTC as an MRI contrast agent, MRI images of tumor-bearing mice were taken before and after injection. SPPTC enhanced the contrast of the MR signal intensity by up to 54% at the tumor site, while it was measured to be only 16% in the SPPC-treated group. This was explained by the accumulation of non-targeted NPs at the tumor via the EPR effect.

Ferritin is the natural iron storage protein possessing a cagelike structure and nanometer size (around 10 nm), with an affinity to the TfR type 1 (TfR1). Apoferritin (APF) is the version of ferritin that contains no iron but has the same targeting ability [139]. Embedding melanin NPs (MNPs) and ferric ions into the cavity of APF were used to construct an efficient nanoplatform, AMF, for in vivo multimodality imaging (PET/MRI/PAI) of colon cancer [140]. The MNPs possessed excellent chelating ability for metal ions (Fe^{3+}, $^{64}Cu^{2+}$) that can be used for MRI and PET and had suitable optical characteristics to be used for PAI. The targeted AMF NPs exhibited higher cellular uptake in HT-29 cells, which had high TfR1 expression compared with HepG2 cells, with lower TfR1 expression.

It was concluded that AMF increased the PET signal intensity 4 h post injection in HT-29 tumor-bearing mice compared with controls. Similar results were achieved with MRI, and the relaxivity value of AMF was two times higher than for controls. Using PAI imaging with 500 µg/mL (based on MNP concentration), the PAI signal of AMF was twofold higher than MNPs, Fe-PEG-MNPs, and AMF without Fe [140]. These data suggested that TfR could be a target for future targeted cancer imaging. Zhu et al. developed a one-step method for the preparation of holo-Tf-indocyanine green (holo-Tf-ICG) nanoassemblies for fluorescence and PA dual-modal imaging and PTT of glioma. The nanoassemblies are formed by hydrophobic interaction and hydrogen bonds between holo-Tf and ICG, which exhibit excellent active tumor targeting and high biocompatibility. A brain tumor with a highly expressed Tf receptor can be clearly observed with holo-Tf-ICG nanoassembly bases on FL and PA dual-modal imaging in subcutaneous and orthotopic glioma models. Under NIR laser irradiation, the holo-Tf-ICG nanoassemblies accumulated in tumor regions can efficiently convert laser energy into hyperthermia for tumor ablation. The novel theranostic nanoplatform holds great promise for the precision diagnosis and treatment of glioma [141]. Physical measurement of tumor volume reduction is the most commonly used method used to assess tumor progression and treatment efficacy in mouse tumor xenograft models, but the detection of tumor size changes can require repeated drug dosing and tumor measurements for several weeks. However, 18F-FDG PET imaging of altered glucose metabolism can be a more sensitive tool for early cancer detection/diagnosis as well as treatment assessment; cancer cells are known to have abnormally increased cellular metabolism that can be inhibited by drug treatment. To illustrate this, Cheih et al. [142] used HCT-116 human colorectal tumor xenografts in nu/nu mice with sorafenib treatment, a clinically approved tyrosine protein kinase inhibitor. Treatment is known to inhibit PDGFR and VEGFR as well as the RAF kinases that regulate energy metabolism in tumors. Sofie G8 PET imaging of treated mice using ^{18}F-FDG revealed a significant drop in tumor metabolism with as little as 2−3 days of treatment, a time during which there is typically little or no effect on tumor size. This approach was relatively low throughput and required special procedures to accommodate the use of radioactivity but offered the option of daily imaging of tumor status. They also explored alternative optical imaging approaches that could offer higher through-put imaging as well as the potential for multiplex imaging. There is no fluorescent equivalent of ^{18}F-FDG, so they focused on bombesin- and transferrin receptors as potentially useful biomarkers for drug-induced inhibition in tumor metabolism. Bombesin receptors are upregulated in a variety of tumors and are important in energy metabolism and tumor growth. The rapid recycling kinetics also make this receptor highly sensitive to cellular metabolic changes. Transferrin receptors are also upregulated in most tumors and provide critical iron transport function vital for their increased enzymatic, proliferative, and metabolic requirements. They used targeted NIRF imaging probes, BombesinRSense 680 (BRS-680) and Transferrin-Vivo 750 (TfV-750), to monitor changes in receptor expression in HCT-116 tumor xenografts during the course of sorafenib treatment. Interestingly, both BR-680 and TfV-750 FLI on the IVIS Spectrum CT yielded data quite similar to our results using ^{18}F-FDG PET; reduction in these probes can be measured as early as 48−72 h in the absence of a significant reduction in tumor size/viability. As expected, both PET and FLI were also highly effective at imaging sorafenib effects 7−8 days later (3−4 days following a 5-day sorafenib treatment regimen), with datasets in good agreement with physical measurements of changes in tumor size. These results suggest that BRS-680 and TfV-750 can serve as fluorescent surrogates for ^{18}F-FDG PET both in measuring early metabolic changes and ultimate therapeutic outcomes following cancer treatment. Goswami et al. developed transferrin (Tf)-templated luminescent blue copper nanoclusters (Tf-Cu NCs) are synthesized. They are further formulated into spherical Tf-Cu NC−doxorubicin nanoparticles (Tf-Cu NC−Dox NPs) based on electrostatic interaction with doxorubicin (Dox). The as-synthesized Tf-Cu NC−Dox NPs are explored for bioimaging and targeted drug delivery to delineate high therapeutic efficacy. FRET within the Tf-Cu NC−Dox NPs exhibited striking red luminescence, wherein the blue luminescence of Tf-Cu NCs (donor) is quenched due to absorption by Dox (acceptor). Interestingly, blue luminescence of Tf-Cu NCs is restored in the cytoplasm of cancer cells upon internalization of the NPs through overexpressed TfR present on the cell surface. Finally, the gradual release of Dox from the NPs led to the generation of its red luminescence inside the nucleus. The biocompatible Tf-Cu NC−Dox NPs displayed superior targeting efficiency on TfR overexpressed cells (HeLa and MCF-7) compared to the cells expressing less TfR (HEK-293 and 3T3-L1). The combination index revealed synergistic activity of Tf-Cu NCs and Dox in Tf-Cu NC−Dox NPs. In vivo assessment of the NPs on TfR-positive Dalton's lymphoma ascites-bearing mice revealed significant inhibition of tumor growth that rendered prolonged survival of the mice [143].

5.2 Folate receptor targeting

The vitamin FA is transported into cells through receptor-mediated endocytosis mediated by the folate receptor (FR), which is overexpressed in cancer CMs compared with normal cells [144]. Different fluorescent nanomaterials such as semiconductor QDs [145], carbon dots (CDs) [146], and small-molecule organic dyes [147] have been decorated with FA to bind to cancer cells in vitro and in vivo. Liu et al. [148] reported the fabrication of a turn-on green fluorescent probe based on FA-modified CDS (FA-CDS) prepared by hydrogen bonding to detect FR-positive cancer cells. The fluorescence intensity of CDs at 520 nm was gradually reduced by increasing the FA concentration, indicating that FA could quench the fluorescence of the CDs. Due to the weak interaction between FA and CDS, when FA binds to the FR, it detaches from the surface of the CDs, resulting in the recovery of the CD fluorescence. Thus, higher concentrations of FR, as found in tumor cells, resulted in stronger fluorescence intensity. There was no significant fluorescence when normal cells were treated with FA-CDs. FA was also conjugated to rhodamine B-labeled poly(propylene fumarate)-co-poly(lactic-co-glycolic acid)-co-poly(ethylene glycol) NPs (PPF-PLGA-PEG-RhB-FA NPs) to track the NPs in both normal osteoblast MC3T3 cells and HeLa cancer cells [149]. In the normal cells, there was no significant difference between the fluorescence intensity of PPF-PLGA-PEG-RhB-FA and PPF-PLGA-PEG-RhB NPs. On the other hand, FA-conjugated NPs showed significantly higher fluorescence in cancer cells. Another study incorporated FA onto the surface of dye-loaded silica NPs as optical nanoprobes for in vitro and in vivo imaging. Depending on the variation in FR expression among the cell lines, their uptake for FA-conjugated silica NPs was different. In vivo imaging indicated that the targeted NPs preferentially accumulated at the site of pancreatic tumor-bearing mice, and there were either weak signals or no signals detected at 24 h and 96 h post injection, respectively. With the exception of the liver, there was no observable fluorescence in the brain, kidney, heart, and spleen, demonstrating good tumor specificity and targeted biodistribution of FA-conjugated silica NPs [150].

FA has also been used as a targeting moiety for many years using CT and MRI imaging modalities. In one such study, FA-linked polyethyleneimine-entrapped gold NPs (FA-Au PENPs) were prepared for tumor CT imaging [151]. Unlike nontargeted Au PENPs, the tumor-targeting ability of FA-Au PENPs via the FR was confirmed by confocal and ICP-OES. For targeted tumor CT imaging, the tumor-bearing mice treated with FA-Au

PENPs showed an obvious enhancement in CT contrast 5 h post injection, with much higher CT values than nontargeted probes. In addition, 1 month later, H&E staining demonstrated that there were no histological changes in the liver, lungs, spleen, kidney, or heart of the mice, which indicated good in vivo biocompatibility of the FA-Au PENPs.

Jin et al. synthesized an FR-mediated dual-targeting drug delivery system to improve the tumor-killing efficiency and inhibit the side effects of anticancer drugs. They designed and synthesized an FR-mediated fluorescence probe (FA-Rho) and FR-mediated cathepsin B-sensitive drug delivery system (FA-GFLG-SN38). FA-GFLG-SN38 is composed of the FR ligand (FA), the tetrapeptide substrate for cathepsin B (GFLG), and an anticancer drug (SN38). The rhodamine B (Rho)-labeled probe FA-Rho is suitable for specific fluorescence imaging of SK-Hep-1 cells overexpressing FR and inactive in FR-negative A549 and 16-HBE cells. FA-GFLG-SN38 exhibited strong cytotoxicity against FR-overexpressing SK-Hep-1, HeLa, and Siha cells, with IC_{50} values of 2–3 µM, but had no effect on FR-negative A549 and 16-HBE cells. The experimental results show that the FA-CFLG-SN38 drug delivery system they proposed can effectively inhibit tumor proliferation in vitro, can be adopted for the diagnosis of tumor tissues, and provides a basis for effective tumor therapy [152]. In another study, Zhang et al. synthesized FA-modified iron oxide (Fe_3O_4) NPs. The in vitro T_2-weighted MR effect of the FA-modified Fe_3O_4 NPs on H460 lung carcinoma cells was evaluated using a 1.5 T MRI machine—the MR signal intensity of the cells showed a significant decrease as a function of Fe concentration, and the obtained images were much darker than those of the same cells treated with FA and nontargeted NPs. Moreover, at equal Fe concentrations, FR-positive cells absorbed more of the FA-modified Fe_3O_4 NPs compared with the FR-negative cells. MRI of H460 tumor-bearing mice injected with FA-modified Fe_3O_4 NPs at different time points was performed. There was a significant reduction in T_2 signal intensity of H460 tumors at 0.85 h post injection [153]. Overall, FR may hold great promise as a target for directed tumor imaging in the future. Kadian and coworkers were first to report a facile in situ synthesis of folic acid-conjugated sulfur-doped graphene QDs (FA-SGQDs) through simple pyrolysis of citric acid (CA), 3-mercaptopropionic acid (MPA), and FA. The as-prepared FA-SGQDs were extensively characterized to confirm the synthesis and incidence of FA molecules on the surface of SGQDs through advanced characterization techniques. Upon excitation at 370-nm wavelength, FA-SGQDs exhibited blue fluorescence with an emission band at 455 nm.

While exhibiting relatively high quantum yield ($\sim 78\%$), favorable biocompatibility, excellent photostability, and desirable optical properties, the FA-SGQDs showed suitability as a fluorescent nanoprobe to distinguish the FR-positive and FR-negative cancer cells. The experimental studies revealed that FA-SGQDs aptly entered into FR-positive cancer cells via a nonimmunogenic FR-mediated endocytosis process. Additionally, the FA-SGQDs exhibited excellent free radical scavenging activity. Hence, these FA-SGQDs hold high promise to serve as efficient fluorescent nanoprobes for the prediagnosis of cancer through targeted bioimaging and other pertinent biological studies [154].

5.3 Epidermal growth factor receptor targeting

The tyrosine kinase EGFR is a 170 kDa transmembrane glycoprotein, which is activated by binding to endogenous ligands of the EGF family. EGFR plays a critical role in cell proliferation, division, inhibition of apoptosis, and angiogenesis, upon activation after internalization via clathrin-mediated endocytosis [155]. The overexpression of EGFR in diverse kinds of malignant tumor cells has been demonstrated [156]. With a high affinity for EGFR ($K_d = 2$ nM), EGF proteins trigger cell proliferation in tumor cells [157]. Therefore, EGFR can be used for targeting cancer cells, using nanoplatforms that have been decorated with EGF proteins or EGFR antibodies for therapeutic and diagnostic applications. Moreover, anti-EGFR antibodies inhibit cell proliferation and trigger cell apoptosis by blocking the activation of EGFR. One study used cetuximab-800CW (anti-EGFR probe) as a fluorescent tracer for ex vivo colonoscopy using an NIR endoscopy platform [158]. The EGFR expression was about 51%−69% higher in 78 low-grade dysplastic adenomas than in normal colon crypts and could be a promising tool for molecular-guided endoscopy. Gao et al. [159] encapsulated the t-BuPITBT-TPE fluorophore within DSPE-PEG NPs decorated with humanized mAb C225 (t-BuPITBT-TPE-C225 NPs) and used this complex for targeted imaging of EGFRs overexpressing non-small-cell lung cancer cells. The t-BuPITBT-TPE-C225 NPs were effectively internalized into EGFR overexpressing HCC827 cells showing a strong red fluorescence compared with only a very weak fluorescence in H23 cells, which express a significantly lower amount of EGFR on their surface.

Recently, multispectral optoacoustic tomography (MSOT) has been used to detect EGFR overexpression in orthotopic pancreatic xenografts, using an NIR EGF-conjugated CF-750 fluorescent probe [160]. Because MSOT is based on the photoacoustic features of the targeted tissue, it is not limited by photon scattering, resulting in high-resolution tomographic images. The specificity and bioactivity of the probe were investigated in different cell lines, including S2VP10L and MiaPaCa-2 cells, with high and low EGFR expression, respectively. After MSOT imaging of S2VP10L–tumor-bearing mice, the EGF-conjugated CF-750 fluorescent probe showed the highest accumulation within the tumor 6 h post injection, with an average of 318 MSOT signal units. However, in mice implanted with MiaPaCa-2 tumors, the MSOT signal was only <10 MSOT signal units. These results indicate good binding and bioactivity of the EGF-conjugated CF-750 probe to EGFR in S2VP10 pancreatic tumor cells and the ability of MSOT to detect the biodistribution of fluorescent dyes in living tissue.

Wang et al. [161] prepared a novel theranostic agent based on PEGylated SPIONs modified with anti-EGFR (Cetuximab) (anti-EGFR-PEG-SPIONs) for MRI and MR-guided focused ultrasound lung cancer surgery. They used this platform to address some limitations, such as the low sensitivity of MRI for visualization of small tumors and the poor efficiency

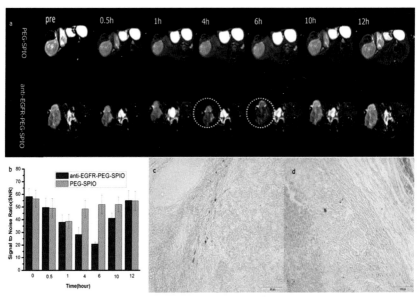

Figure 4.14 *Tumor imaging with anti-EGFR-PEG-SPIONs.* T_2WI MRI images (A) and SNR (B) of tumor after injection of 0.1 mL targeted and nontargeted contrast agents at different time points (0.5, 1, 4, 6, 10, 12 h). The mean T_2-weighted signal intensities were measured for each tumor. The relative signal-to-noise ratios (SNRs) were calculated. Prussian blue staining of tumor tissues after 6 h injection of (C) anti-EGFR-PEG-SPIONs and (D) PEGylated SPIONs. *(Reproduced with permission Wang Z, et al. Active targeting theranostic iron oxide nanoparticles for MRI and magnetic resonance-guided focused ultrasound ablation of lung cancer. Biomaterials 2017;127:25–35. Copyright 2017, Elsevier).*

of in vivo ultrasonic energy deposition. In vivo, MRI employed two groups of H460 lung tumor-bearing nude rats injected with anti-EGFR-PEG-SPIONs and PEGylated SPIONs (Fig. 4.14). At 4 h post injection of the targeted NPs, the T_2 signal-to-noise ratio (SNR) showed a significant decrease at the tumor site compared with only a slight decrease with nontargeted NPs. This was explained by the targeting ability of anti-EGFR-PEG-SPIONs to the overexpressed EGFR on the H460 lung cancer cells. They also employed Prussian blue staining to confirm the targeted contrast agent had a higher deposition in tumor tissue than the nontargeted NPs (Fig. 4.14C).

5.4 Glucose transporter targeting

Both primary and metastatic cancer cells consume higher volumes of glucose to provide themselves with energy, which is required for their rapid proliferation [162]. This increased glucose requirement results in upregulation of the glucose transporter (Glut) (e.g., Glut-1 and Glut-3) on the surface of cancer cells. This observation led to clinical imaging being revolutionized by the invention of a new imaging approach termed FDG PET. PET is able to produce high-resolution anatomical images based on the preferential uptake of glucose into cancer cells compared with normal, employing the glucose analog fluorodeoxyglucose labeled with the PET isotope, 18-fluorine (half-life 10 min) [163]. However, this technique is not considered very specific for cancer because other biological mechanisms (like inflammation) result in higher metabolic uptake of glucose. Therefore, glucose imaging probes must distinguish between cancer and inflammation.

Inspired by PET, many researchers have taken advantage of the abnormal expression of Glut (as well as increased glucose metabolism) as a hallmark of cancer to trace and image tumors. To this end, glucose (and its many derivatives) has been attached to different nanoplatforms, and its ability to target Gluts on cancer cells has been evaluated. Dreifuss et al. [164] employed glucose-functionalized gold nanoparticles (GF-GNPs) as a metabolically targeted CT contrast agent. They hypothesized that the cellular uptake of larger-sized GF-GNPs via GLUT-1 was unlikely compared with the facile uptake of small glucose molecules. This led them to propose that GLU-1 induced a biological cascade that eventually resulted in the increased uptake of GF-GNPs, probably via endocytosis. Based on the CT images, the GF-GNPs could differentiate between cancer and inflammation in a mouse model that combined both tumor and inflammation at different sites, possibly because of differences in the vasculature of the different pathologic conditions. Singh et al. [165] compared the mechanism of internalization between BSA-coated gold nanoclusters (BSA-AuNCs) and glucose-coated gold nanoclusters (Glu-AuNCs) in human epithelial carcinoma (A431) cells and the human keratinocyte cell line (HaCaT) as examples of cancerous and noncancerous cells, respectively. Based on fluorescence imaging, Glu-AuNCs were internalized by A431 cells via Glut-1 receptors, while there was significantly lower internalization by HaCaT cells. Likewise, BSA-AuNCs showed significantly higher cellular uptake in A431 cells than HaCaT cells; however, this internalization was dependent on the CM potential, which affected the electrostatic interaction between cells and NPs. They further investigated the level of Glut-1 protein expression in both cell lines. As expected, the expression level of Glut-1 was 40% higher in A431 than in HaCaT cells.

Recently, Zhao et al. [28] reported the preparation of a novel dual-stimulus responsive nanoprobe for in vivo tumor-specific image-guided photothermal therapy. The nanoprobe called Pep-Acy/Glu@AuNRs consisted of four constituents: (1) gold nanorods (AuNRs) as the basic structure, photothermal therapy agent, and ultraefficient fluorescent quencher; (2) an asymmetric fluorescent cyanine dye (Acy) which served as a tumor-specific imaging probe with pH-responsive NIR absorption and fluorescence; (3) MMP-specific peptide (Pep) acting as a linker between the AuNRs and Acy; and (4) glycosyl residues (Glu) on the surface of AuNRs providing active tumor-targeting ability. In the presence of MMP-13 at pH 6.0, there was intense fluorescence due to the detachment of Acy from the AuNRs because Pep was cleaved by the MMP enzyme and pH-sensitive activation of Acy into its acidic fluorescent form. Pep-Acy/Glu@AuNRs were not fluorescent either at pH 7.4 or in the absence of MMP-13. The in vitro cell internalization revealed that at pH 6.0, SCC-7 cells incubated with Pep-Acy/Glu@AuNRs showed 2.7-fold higher fluorescence intensity than the Glut-blocked SCC-7 cells, indicating the critical role of Glut in the cellular uptake of NPs via Glu interaction with Glut. They then explored the ability of Pep-Acy/Glu@AuNRs to allow in vivo precision tumor-targeting imaging in SCC-7 tumor-bearing nude mice (Fig. 4.8). The fluorescence signal was intense and lasted for up to 12h in the active tumors (R-tumor). In contrast, by manipulating the TME using either an MMP inhibitor (for Group A) or NaHCO$_3$ (for Group B), there was almost no fluorescence signal detected in either group. These results demonstrated the various features of the designed nanoplatform, including dual-stimuli responsivity, accuracy, tumor targeting via the microenvironment, and overexpression of Glut within the tumor cells. Cheng et al. designed and synthesized a novel C1-type glucose conjugated deep-red emissive probe, Glu-1-O-DCSN, for in vitro real-time cellular glucose uptake imaging and in vivo glucose uptake and utilization tracing. The significant accumulation in the tumor promises various applications of Glu-1-O-DCSN as an indicator for anticancer drug screening and assessment and a clinical tracer in intraoperative tumor-targeted imaging approaches for surgical navigation [166]. Probes with long lifetime fluorescent emission are highly desirable for time-resolved fluorescence imaging to eliminate the interference from short-lived autofluorescence. Aromatic-imide-based thermally activated delayed fluorescence material TADF compound with high external quantum efficiency was encapsulated with Glucose-PEG2000-DSPE amphiphilic copolymer to prepare nanoprobe micelles. Glucose-PEG2000-DSPE TADF (GPDT) Micelles had good biocompatibility, biodegradability, less toxicity, and they could be efficiently transported by overexpressed GLUT1 on the tumor CM, then efficiently targeted lysosome of HepG2 cells. GPDT Micelles had great clinical application and research development value and be suitable for the application and popularization in fluorescence imaging technology, especially time-resolved fluorescent imaging in living cells [167]. Rasouli et al. proposed a new nanostructure to investigate the targeting of tyrosine-conjugated ultrasmall superparamagnetic iron oxide nanoparticles (USPIONs) as a targeted nanocontrast agent for application in molecular MRI of breast cancer. Recently, studies demonstrated that L-type amino acid transporters (LAT1) are highly expressed in breast cancer cells. Thus, LAT1 targeting via tyrosine as a LAT1 substrate could improve the sensitivity and specificity of this nanosized contrast agent. To achieve this goal, USPIONs were conjugated to tyrosine and characterized using DLS and FT-IR. The cell viability was measured in different concentrations of NPs in breast cancer cells (MDA231, MCF7,4T1) and control cell line (normal kidney cells; HEK293) with the MTT assay. Cellular uptake was evaluated via Prussian blue staining and inductively coupled plasma-optical emission spectroscopy (ICP-OES) as well as by measuring reduced signal intensity using 3 Tesla clinical MRI. T2 weighted imaging in tumor-bearing bulb/c mice was performed via brain coil and home-built phantom. The particle size and charge of USPIO significantly changed after the conjugation of tyrosine. According to ICP-OES results from the cellular uptake of tyrosine-USPION in the 4T1 cell line and HEK-293 was 48.14% $\pm$ 1.43 and 6.91 $\pm$ 0.21, respectively. The reduction in MRI signal intensity at in vitro studies was higher in the presence of tyrosine-USPION than of USPION. The reduction in MRI signal intensity at in vivo studies was 58.83% in the presence of tyrosine conjugated USPION compared with plain NPs. Biodistribution studies demonstrated that the accumulation of tyrosine-USPIONs was about seven times higher than that of nontargeted USPIONs after 24h. In conclusion, tyrosine-USPIO as a new LAT1 targeted contrast agent with high sensitivity and specificity can be suggested as a good and ideal candidate in breast cancer molecular imaging [168].

5.5 Cathepsin targeting

Cathepsins (Cats) are a group of lysosomal peptidases with a cysteine residue at the active enzymic site. Cats are proteins that degrade the ECM and the basal membrane, which is essential for tumorigenesis. In humans, the Cat family comprises 11 members (Cat B, Cat C, Cat F, Cat H, Cat K, Cat L (Cat L1), Cat L2 (Cat V), Cat O, Cat S, Cat W, and CatZ [Cat X]). The majority of Cats are endopeptidases that cleave peptide bonds within their protein substrates (Cat C and Cat Z do not have any endopeptidase activity). Cat B possesses carboxypeptidase activity, and Cat H possesses aminopeptidase activity. The functional contributions of the cysteine cathepsins to tumor invasion and metastasis are diverse. For example, Cat B, C, K, L, S, V, and X are all expressed in tumor-associated macrophages. Cats B, F, H, K, L, S, V, and X are all expressed in different tumor cells. Increased Cat expression is correlated with poor prognosis in breast (Cat B, L, C, and S), lung (Cat B, H, and S), ovarian cancer (Cat B), pancreatic (Cat B, L, and Z), osteosarcoma (Cat K) and colorectal cancer (Cat B, L, and S) [169,170].

Because Cat B is an attractive target for the detection of tumor metastases, Ryu et al. [171] developed a Cat B-sensitive nanoprobe (Cat B–CNP) consisting of a self-quenched Cat B-sensitive fluorogenic peptide (Gly-Arg-Arg-Gly-Lys-Gly-Gly) probe conjugated onto the surface of tumor-targeting glycol-chitosan NPs. This platform facilitated sensitive and specific visualization of Cat B activity in cells and in vivo tumor models. Cat B–CNP demonstrated the potential to distinguish metastases in three metastatic mouse models, including lung, liver, and peritoneal metastases. Metabolic glycol-engineering with a biorthogonal click reaction has been used to improve the tumor-targeting efficiency of NPs as delivery vehicles for imaging agents or for anticancer drugs. This technique can develop metabolic agents that can create abnormal glycans expressed on the tumor cell surface that can then be labeled with click chemistry approaches.

Shim et al. [172] developed a Cat B-specific metabolic precursor consisting of a Cat B-specific cleavable substrate (Lys-Gly-Arg-Arg (KGRR)) conjugated to triacetylated N-azidoacetyl-D-mannose amine (RR-S-Ac$_3$ManNAz) for the creation of azide-containing glycans on tumor cells. Subsequently, these azido–glycans could be labeled by NIRF dyes containing triple bonds using the click reaction. In vivo imaging results showed this system could be a promising tool for tumor-specific active targeting.

An amino-functionalized metal−organic framework (MOF) could be an efficient delivery vehicle for cell imaging and chemo-PDT. Liu et al. [173] developed a multifunctional MOF nanoprobe loaded with campto-thecin (chemotherapy drug), FA (targeting moiety), and a chlorin (e6) (Ce6)-conjugated Cat B-substrate peptide as the activatable moiety. The MOF probe recognized FR-positive tumor cells, where Cat B activated the release of Ce6 as an imaging agent and a photosensitizer and camptothecin as an anticancer drug.

Targeting cancerous tissue with iodinated CT contrast agents could be improved by taking advantage of enzyme overexpression using activity-based probe (ABP) methodology. A typical ABP includes a recognition element that is a substrate for a tumor-associated protease, a contrast agent, and a "warhead" (usually an electrophile that can form a covalent linkage between the target and the contrast agent). When CT is used for cancer imaging applications, its relatively low contrast requires the use of high concentrations of contrast agents. To overcome this limitation, Gaikwad et al. prepared a new class of iodinated nanoscale ABPs (IN-ABPs) that could enrich the concentration of iodine at the targeted tumor site by covalent attachment in the presence of Cats that are significantly overexpressed in cancer. The IN-ABPs comprised a short targeting peptide sequence selective for specific Cats, an electrophilic moiety that allowed activity-dependent covalent binding and tagged with dendrimers loaded with iodine atoms. IN-ABPs selectively bound to tumors in the presence of recombinant and intracellular Cats B, L, and S. They compared the in vivo biodistribution and tumor accumulation of IN-ABPs bearing 18 or 48 iodine atoms. The result of this study showed the synthetic feasibility and potential utility of ABPs as potent contrast agents for CT of tumors [174]. Employing a similar strategy, Tsvirkun et al. [175] developed nanosized Cat-targeted ABPs for functional CT imaging. Their probe consisted of various sizes of gold NPs with varying ratios of Cat-targeted substrate and PEG. The results showed that GNP-ABPs were a promising tool for enzymatic-based CT imaging.

6. Cancer-specific ligand targeting

The TME is characterized by altered functions in ECM molecules, vascularized stroma, lymphatic networks, and abnormal cell phenotypes. The molecular imaging of specific cancer cell types plays a critical role in tumor detection, as described below. Depending on the type and stage of cancer, different antigens or receptors are overexpressed on the surface of the cancer cells and can be used in ligand-mediated targeted tumor imaging. Ligand targeting increases interactions between the contrast agent and the targeted cells and enhances the cell internalization of the agent without altering the overall biodistribution [176]. There are two types of cancer-specific ligands: (1) serum markers that are mostly used for monitoring of patients with already diagnosed disease, predicting response to therapy, and determination of prognosis; (2) markers that exist within the tumor tissue (cell surface) and are used for molecular imaging and detection of cancer. Because serum markers have low sensitivity and are not useful for early detection, we have concentrated on the tumor tissue markers in most common cancer types with an annual incidence of 40,000 cases and high mortality rates according to NCI data (https://www.cancer.gov/types/common-cancers).

6.1 Targeted breast cancer imaging

Breast cancer is a multifaceted and heterogeneous disease with a high worldwide mortality rate [177]. Based on NCI data, breast cancer was the most common type of cancer diagnosed in 2018, with 266,120 new cases and 40,920 deaths [178]. Specific breast tumor tissue markers used in diagnosis and therapy, such as estrogen receptor (ER), progesterone receptor (PR), hormone receptor (HR), HER2 gene (also known as c-erbB-2 or neu), urokinase plasminogen activator (uPA), and plasminogen activator inhibitor 1 (PAI-1) are covered in Table 4.3. These are the most promising biomarkers in lymph node-negative breast cancer. However, only a select few of these markers have been clinically used for imaging. Depending on changes in the levels of ER, PR, HR, and HER2, breast cancer is often characterized into subtypes: luminal A ($ER^+/PR^+/HER2^-$), luminal B ($ER^+/PR^+/HER2^+$), HER2 overexpressing ($ER^-/PR^-/HER2^+$) and TNBC ($ER^-/PR^-/HER^-$) [179–182]. Luminal tumors ($\sim70\%$ of invasive breast cancers) respond to hormonal therapy, and the HER2 overexpressing subtype responds to targeted antibody therapy. TNBCs are more aggressive and difficult to treat but may respond to

Table 4.3 Summary of imaging platforms used for the detection of a range of human cancers.

Biomarker (target)	Ligand	Imaging platform	Status	Refs
Breast cancer				
gC1q receptor p32 protein	Peptide, CGNKRTRGC (LyP1)	Bi$_2$S$_3$-LyP-1	In vitro/ in vivo	[266]
Phosphatidylserine (PS)	Monoclonal antibody (mAb), PGN635	PGN-L-IO/DiR	In vitro/ in vivo	[267]
Urokinase plasminogen activator receptor (uPAR)	Amino-terminal fragments (ATF)	NIR830-ATF-IONP	In vivo	[268]
Epithelial cell adhesion molecule (EpCAM)	Aptamer	Apt-QD-nut-NPs	In vitro/ in vivo	[269]
Macrophage mannose receptor (MMR; CD206)	Anti-CD206 Ab	Dye-anti-CD206	In vitro/ in vivo	[270]
CD44	Hyaluronic acid	HA-dOG-PTX-PM	In vivo	[271]
HER2	Trastuzumab	^{89}Zr-trastuzumab	Clinical trial	[272]
Gastrin releasing peptide (GRP) receptors	Bombesin (BBN)	DSPION–BBN	In vitro/ in vivo	[273]
Chemokine receptor CXCR4	Pentixafor	68Ga-Pentixafor	In vitro/ in vivo	[274]
Sodium iodide symporter (NIS)-mediated nuclear reporter	I-124	I-124	In vitro/ in vivo	[275]

CD146	YY146 Ab	64Cu–NOTA–YY146	In vitro/in vivo [202]
CD38	IgG Ab	89Zr–Df–IgG	In vitro/In vivo [276]
Subtype somatostatin receptor 2 (SSTR2)	PA1 peptide	68Ga–DOTA–PA1	In vitro/In vivo [277]
CD30	Brentuximab vedotin (BV)	89Zr–Df–BV	In vitro/In vivo [278]
Monoclonal antibody, h173	mAb, h173	64Cu–DOTA–h173	In vitro/In vivo [279]
Colorectal cancer			
Translocator protein (TSPO)	[18F] FEPPA (N-acetyl-N-(2-[18F]-fluoroethoxyben-zyl)-2-phenoxy-5-pyridinamine)	TSPO-PET tracer [18F] FEPPA	In vitro [280]
VEGFR–1 and NRP–1 (neuropilin–1)	CPQPRPLC peptide	[99mTc] Tc-HYNIC-D(LPR) peptide for SPECT imaging	In vitro/In vivo [281]
Metastatic SW620 membrane protein	SW620-specific DNA aptamer	Aptamer XL-33	In vitro/Ex vivo [282]
EpCAM	Anti EpCAM mAb	UCNP@SiO$_2$ Rose Bengal (RB)-Linker–Protein G (LPG)-Anti EpCAM	In vitro [283]
Colon cancer secreted protein-2 (CCSP-2)	Anti–CCSP-2 antibody	Anti–CCSP-2 antibody-FPR–675	In vitro/In vivo [284]
Hexosaminidase	Probe for β-galactosidase	Hexosaminidase (HMRef-βGlcNAc)	In vitro/Ex vivo [285]

Continued

Table 4.3 Summary of imaging platforms used for the detection of a range of human cancers.—cont'd

Biomarker (target)	Ligand	Imaging platform	Status	Refs
Carcinoembryonic antigen (CEA)	Anti-CEA Ig G	IgG-conjugated fluorescent nanoparticles	In vivo	[286]
Claudin-1	RTSPSSR peptide	Cy5.5-a GGGSK linker-peptide	In vivo	[287]
Delta-like ligand 4 (Dll4)	Dll4 mAb (61B)	61B-DOTA-64Cu PET probe	In vitro/ In vivo	[288]
Prostate cancer				
PSMA	HBED (PSMA-targeted probe)	^{68}Ga-HBED-CC PSMA PET	In vivo (PCa patients)	[289]
PSMA	Anti-PASMA Ab	QD- PEG -PSMA Ab	In vivo	[290]
Gastric-releasing peptide receptors (GRPRs)	BBN	BBN-conjugated Cy5.5 - N-acetyl histidine - glycol chitosan NPs	In vitro/ In vivo	[291]
PSMA	xPSM-A9 and xPSM-A10 (RNA aptamers)/RGD/Ab	G4.5 PAMAM dendrimer- iron oxide	In vitro/ In vivo	[292]
SPARC glycoprotein	M13 filamentous bacteriophage	M13-SBP-MNP	In vitro/ In vivo	[293]
Androgen receptor (AR)	Peptide SP204	Superparamagnetic iron oxide nanoparticles (SPIONs– SP204	In vitro/ In vivo	[223]
Prostate stem cell antigen (PSCA)	Anti-PSCA mAb	GO-DEN(Gd-DTPA)-mAb	In vivo	[294]
GRPRs	BBN analogue (named Scheme 1) based on JMV594 peptide	^{68}Ga-NODAGA-SCH1	In vitro/ In vivo	[295]

Target	Agent	Probe	Application	Reference
Insulin-like growth factor 1 (IGF1) receptor	Ab clone (1A2G11)	^{64}Cu-NOTA-1A2G11	In vitro/In vivo	[296]
Urokinase-type plasminogen activator receptor (uPAR)	AE105 Ab	^{64}Cu-DOTA-AE105	In vitro/In vivo	[297]
GRPRs	Peptides NOTA-BBN2	^{68}Ga-NOTA-BBN2	In vivo	[298]
Pancreatic cancer				
CD47	iExosomes	iExosomes	In vitro/In vivo	[299]
BBN receptors	BBN peptide	**BBN**-CLIO(Cy5.5)	In vivo	[300]
Toll-like receptor 2 (TLR2)	TLR2 agonists	TLR2 agonists-IR800CW	In vitro	[301]
IGF1 receptor	IGF1	IGF1-IONP-Dox	In vitro/In vivo	[231]
Claudin-4	Anti-Claudin-4	QDs- anti-Claudin-4	In vitro	[302]
uPAR	Amino-terminal fragment (ATF) peptide	ATF-IONP-(GFLG)-Gem	In vitro/In vivo	[303]
KRAS2 mRNA	KRAS2 PNA-D (Cys-Ser-Lys-Cys)	[111In]DOTAn-poly(diamidopropanoyl)m-KRAS2 PNA-D(Cys-Ser-Lys-Cys)	In vivo	[304]
Neurotensin (NT) receptors	NT	Aluminum-^{18}F-NOTA-NT	In vivo	[305]
CA19.9	Anti-CA19.9 Ab 5B1	^{89}Zr–5B1	In vivo	[306]
Bladder cancer				
CD44	CD44v6 Ab	Liposomal nanoprobe	In vivo	[307]
CD47	CD47 Ab	Antibody-functionalized SERS nanoparticles	In vitro/In vivo	[244]

Continued

Table 4.3 Summary of imaging platforms used for the detection of a range of human cancers.—cont'd

Biomarker (target)	Ligand	Imaging platform	Status	Refs
Prostate stem cell antigen (PSCA)	Anti-PSCA mAb	QD-PSCA	In vitro	[308]
Carbonic anhydrase IX (CAIX)	Anti-CAIX Ab	Anti-CAIX-Qdot625	In vitro/ ex vivo	[309]
Argininosuccinate synthetase 1	Fluoro-L-thymidine (FLT)	[^{18}F]-fluoro-L-thymidine (FLT)	In vitro/ In vivo	[310]
CD47	CD47 Ab	Anti-CD47–FITC	In vivo	[242]
IL-5Rα	mAb A14	^{64}Cu-A14	In vitro/ In vivo	[311]
Brain cancer				
Transferrin receptor TfR & HER2	Transferrin peptide (Tf$_{pep}$) Poly (β-L-malic acid) polymeric nanoimaging agents (NIAs)	Tf$_{pep}$-Au NPs Gadolinium-DOTA- poly (β-L-malic acid) polymeric nanoimaging agents (NIAs)	In vitro In vitro/ In vivo	[312] [313]
EGFR & CD105	Denoted as Bs-F (ab)$_2$]	^{64}Cu-NOTA-Bs-F (ab)$_2$	In vitro/ In vivo	[314]
Lipoprotein receptor-related protein (LRP) receptors & αVβ3	Angiopep-2 peptides and cyclic [RGDyK] peptides	PAMAM-G5 dendrimer	In vitro/ In vivo	[315]
LRP-1	ANG	ANG –Tetrahedral DNA nanostructures (TDNs)	In vivo	[316]
LRP-1	Angiopep-2 (ANG, TFFYGGSRGKRNNFKTEEY)	ANG/PLGA/DTX/ICG	In vitro/ In vivo	[317]
CD146	YY146, a high affinity anti-CD146 mAb	^{89}Zr-Df-YY146	In vitro/ ex vivo	[318]

Ovarian cancer

IL–16	anti-IL–16 Ab	Microbubbles– anti–IL–16	In vivo	[319]
Death receptor 6 (DR6)	Anti-chicken DR6 Ab	Microbubbles-anti-chicken DR6 Ab	In vivo	[320]
CD276	Anti-CD276 Ab	Microbubbles– anti-CD276 Ab	In vivo	[321]
HER2	Trastuzumab	^{89}Zr-trastuzumab	In vivo	[322]
HER3	HER3–antibody RG7116	^{89}Zr-RG7116	Phase I	[323]
Mesothelin	Anti-Mesothelin nanobody (NbG3a)	NbG3a-IONP	In vivo	[324]
HER–2	anti-HER–2 mAb	Optical viral ghosts (OVGs)—ICG-anti-HER–2	In vitro/In vivo	[325]
Cyclooxygenase-1 (COX-1)	P6	[18F]-P6	In vivo	[326]
Poly (ADP–ribose) polymerase (PARP)	Fluorthanatrace (FTT)	[^{18}F] FTT	In vitro/in vivo	[327]

chemotherapy. Breast cancer is classified into five stages (0, I, II, III, and IV) with different marker expression, biology, and therapeutic responses. Hence, a full understanding of breast cancer subtypes is important for the success of treatment outcomes.

A recent report described a CD44-targeted nanomicellar payload delivery platform for selective tumor-specific imaging and therapy of TNBC [183]. Several mAbs, e.g., trastuzumab (Herceptin) and pertuzumab, or small-molecule tyrosine kinase inhibitors such as neratinib and lapatinib, have been utilized for targeting $HER2^+$-overexpressing breast tumors [184]. A range of specific molecules, such as "protein–phosphatase 2A-regulatory molecule (B) 55 β-subunit" (PP2A-B55β), P7170 (synthetic inhibitor), IL-15 receptor and it is α subunit (IL15RA), and progesterone receptor (PgR) have been used for clinical targeting and inhibition of TNBC subtypes.

Radiolabeled Fluoroestradiol (^{18}F-FES) PET/CT imaging has been utilized for primary ER^+ breast cancer detection, evaluation of metastases, and monitoring response to endocrine therapy. FES is an estrogen hormone with the affinity to bind to ERα and is used as a targeted contrast agent for ^{18}F-FES PET/CT imaging [185]. Heat shock proteins (Hsps) can act as a breast tumor marker and have been targeted using multifunctional NPs based on perfluoropolyether (PFPE)-conjugated peptide aptamers that specifically bind to Hsp70 and act as fluorescent and MRI contrast agents. The in vivo results demonstrated the platform possessed high tumor accumulation with a specific affinity to Hsp70. These peptide aptamers could effectively target the TME (surface of the tumor cells) and the interior of the tumor cells [186].

Neu or HER2 is a proto-oncogene that is overexpressed in up to 30% of breast cancers. Kievit et al. [187] developed an imaging probe constructed from superparamagnetic iron oxide nanoparticles (SPIONs) coated with copolymer chitosan grafted PEG and then conjugated with an anti-neu antibody. MR imaging demonstrated that the probe could target neu receptors in vitro and in vivo in a transgenic mouse model. Furthermore, this probe recognized and tagged spontaneous micrometastases in the liver, bone marrow, and lungs of tumor-bearing mice.

The overexpression of the chemokine receptor (CXCR4) plays an important role in breast cancer cell proliferation, invasion, and metastasis. One study developed ^{64}Cu-doped gold nanoclusters conjugated to AMD3100, a ligand that specifically binds to CXCR4. The ^{64}Cu−AuNC-AMD3100 was used for the detection of lung metastasis in a mouse model bearing 4T1 metastatic breast cancer. The PET imaging results showed that the contrast agent had excellent affinity and sensitivity for targeting CXCR4, both in the early stages of the tumor and in micrometastases in the lungs [188]. Y_1 receptors (Y_1Rs) are also highly overexpressed in human breast cancer and its metastases. Fluorescent nanobubbles (NBs) were fabricated from tetradecafluorohexane and biodegradable photo-luminescent polymers and then conjugated to a PNBL-NPY ligand developed for specific targeting of Y_1Rs both in vitro and in vivo. The results showed PNBL-NPY-modified NBs had good dispersity, biocompatibility, and stability and possessed high affinity and specificity for Y_1Rs [189]. To increase specificity, probe circulation time, and precise targeting, a dual-targeting strategy was described using hybrid GNRs conjugated to Herceptin (HER) and PEG. The imaging results showed good accumulation of the Her-PEG-GNRs in tumors compared with Her-GNR and PEG-GNR tested alone [190].

6.2 Targeted lung cancer imaging

Lung cancer is highly invasive and metastatic, with one of the highest cancer mortality rates worldwide [191]. According to the NCI, an estimated 234,030 new cases and 154,050 deaths from lung cancer were reported in 2018. Lung cancer is a heterogeneous disease and is difficult to diagnose early in many cases. The disease is often diagnosed only in advanced stages (stage III or VI). The lung cancer subtypes include (1) squamous cell lung cancers (SQCLC), which account for approximately 25%−30% of all cases and arise from the main bronchi and spread to the carina; (2) adenocarcinomas (adenoCA), which represent about 40% of all lung cancers and arises from peripheral bronchi; (3) lung cell anaplastic carcinomas (LCAC), which represent about 10% of all lung cancers and lack classic glandular or squamous morphology in the tumor; and (4) small-cell lung cancer (SCLC), which accounts for approximately 10%−15% of all lung cancers and arises from the lung neuroendocrine cells, and disseminates into the submucosal lymphatic vessels and regional lymph nodes without any bronchial invasion. Based on histological data, lung cancer is divided into two classes, with different growth and spread profiles: NSCLC, which consist of adenoCA, LCAC, and SQCLC subtypes, and accounts for approximately 85%−90% of all lung cancers; and SCLC accounting for approximately 10%−15% of lung cancers. All lung cancer subtypes can become multifocal within the part of the lung they first occur (T3), spread throughout the lung of origin (T4), or spread to the contralateral lung (M1) [192−194].

Metastatic lung cancer can be diagnosed in inaccessible sites such as the bone, liver, or brain before any symptoms occur due to the primary lung lesion. Depending on genetic alterations, lung cancer can be classified in several ways, including (1) activation of mutations in proto-oncogenes such as BRAF, MEK, KRAS, PI3K, HER2, FAT2, GPR87, LYPD3, SLC7ALL, and especially EGFR; (2) amplification of proto-oncogenes, such as fibroblast growth factor receptor 1 (FGFR1) and discoidin domain receptor (DDR2) in SQCLC, and MET in adenoCA; (3) gene activation in anaplastic lymphoma kinase, rearranged during transfection (RET), or c-ros oncogene 1 (ROS1); (4) overexpression of miRNAs; (5) inactivation of tumor suppressor genes (TSGs), including RB1, CDKN2A, TP53, PTEN, FHIT, RASSF1A; and (6) increased telomerase activity [195–199].

EGFR is mutated and overexpressed in almost 80% of NSCLC cases. Anti-EGFR Abs have been used as contrast agents in lung cancers. However, Ab production is difficult and costly, and Abs have a rather low tumor penetration due to their large size. Therefore, a short peptide sequence (P75) was introduced as an EGFR-targeting peptide and used for CT/photoacoustic dual-modality image-guided photothermal therapy, Zhao et al. [200] designed P75-modified triangular gold NSs (P75-PEG-TGN). The in vitro and in vivo results showed high affinity to EGFR$^+$ cancer cells and increased accumulation on the tumor cell surface. The cytotoxic T-lymphocyte-associated protein 4 (CTLA-4) is a marker of immune T cells and some lung cancer cells. The Ehlerding group [201] used ^{64}Cu radiolabeled ipilimumab (anti-CTLA-4 mAb) for PET imaging of human NSCLC cells. In vivo results showed the radiolabeled Ab effectively accumulated in CTLA-4$^+$ NSCLC.

Other CD markers (i.e., CD30, CD48, CD146, CD44, and CD133) are overexpressed on the surface of lung cancer cells. Overexpression of CD146 (or MUC18) is associated with metastatic potential and is detectable in 50%–75% of lung cancers. The YY146 mAb was radiolabeled using ^{64}Cu (^{64}Cu–NOTA-YY146) as a targeted contrast agent for in vitro and in vivo PET imaging of CD146+ intrapulmonary metastases of NSCLC cells [202]. Additionally, the delta-opioid receptor (δOR) (a member of the G-protein receptor family) is overexpressed in human lung cancer cells but not expressed in normal lung cells. Cohen et al. [203] described an imaging system using synthetic Dmt-Tic peptide (a δOR antagonist) and IR800 NIR dye that had an excellent affinity for δOR for in vitro lung cancer cell imaging. More examples are provided in Table 4.3.

6.3 Targeted colorectal cancer imaging

Colorectal cancer (CRC) is the second-most and third-most common cancer in the United States for women and men, respectively. More-developed regions of the world have a higher incidence than less-developed regions. Based on the standardized incidence rate (ASRi), the majority of patients with sporadic CRC are >50 years of age, and both genetic factors (e.g., mutations in the DNA mismatch-repair genes, proto-oncogenes, and TSG) and environmental factors (e.g., smoking, alcohol intake, and increased body weight) contribute to the etiology of CRC. Furthermore, epigenetic alterations seem to affect gene expression to trigger changes in benign polyps into malignant tumors [204]. Nowadays, CRC diagnosis relies on an assessment of patient symptoms and is followed by an instrumental approach if needed (i.e., colonoscopy, capsule endoscopy, CT colonography, and measurement of prognostic/predictive biomarkers of CRC). CRC biomarkers can be categorized into diagnostic, pharmaco-logical, predictive, prognostic risk/predisposition, screening, and surrogate response biomarkers [205]. The most important DNA biomarkers are microsatellite instability, aberrant methylation of septin 9 (SEPT9) (a GTPase), mutation of adenomatous polyposis coli, and Kirsten rat sarcoma (KRAS) [206]. CRC (colon and rectal cancers) and their stages are important for the treatment of CRC. Surgery and targeted therapy using mAbs against overexpressed factors (e.g., EGFR, VEGF-A, HER) is the mainstay potentially curative treatment for patients with nonmetastatic tumors (stages I–III), while fusion proteins that target multiple proangio-genic growth factors have been utilized for metastatic CRC (Stage IV) [207]. Beyond conventional imaging modalities, PET-CT is a novel mo-lecular imaging approach employing radiolabeled Abs or Ab fragments to detect CRC overexpressing EGFR [208] (Table 4.3).

If metastatic disease becomes clinically established, the long-term patient outcomes are not favorable, and current imaging rarely detects the early stages of cancer development at either the primary or metastatic sites. NPs have been found to accumulate in tumors in large numbers [209]. CRC diagnosis and treatment could also be improved by employing NPs or nanoprobes. Detection of the small polyp was enabled using a nanobeacon comprising polystyrene NPs with coumarin 6 dyes encapsulated within the core and a surface decorated with poly (N-vinylacetamide) (PNVA) and coated with peanut agglutinin (PNA). These NPs showed high binding affinity to the CRC-associated Thomsen-Friedenreich (TF) antigen. The designed nanobeacon could be used for the clinical detection of hidden polyps, early quantitative detection of CRC, and for distinguishing adenomas and adenocarcinomas from normal colonic tissue [210].

The multiple vibrational modes of NIR emission can be improved by using QDs. Unlike organic dyes, QDs can allow for multiplexed imaging due to the narrow-band emissions. Development of a protease-activatable QD (PbS/CdS/ZnS core/shell/shell) probe emitting in the NIR-II spectral region (PA-NIRQD) showed selective fluorescence activation and a high signal peak in the presence of MMP enzyme activity at tumor sites in a colon cancer mouse model [211]. To overcome poor tissue penetration of the light and the background autofluorescence of traditional fluorescence-based imaging probes, one study used multifunctional silica-based nanocapsules containing two distinct triplet–triplet annihilation upconversion (TTA-UC) chromophore pairs that were then conjugated with TCP (a vasculature-targeting peptide for CRC). The experimental results demonstrated that this platform is bound only to CRC cells with differential-color imaging and greater accumulation at targeted tumor sites and is a promising tool for CRC diagnosis within the heterogeneous TME [212].

Images generated by fluorescent microscopy/endomicroscopy, such as two-photon microscopy (TPM), have high resolution that enables visualization of biological processes (such as cell trafficking and cell–cell interaction). The morphology of biopsies taken from diseased colon could be visualized without fixation and staining. Beack et al. [213] developed a PNA-conjugated hyaluronate with a high affinity to CD44/CD44v6 receptors for colon cancer detection to enable image-guided endoscopic resection of a large colorectal polyp. TPM of rhodamine B fluorescence was used for bioimaging of CRC. Another strategy to improve the detection of smaller or nonpolypoid lesions, which have miss rates of up to 24% during colonoscopy, is to combine advanced imaging technology and targeted molecular probes, preferably using biomarkers that apply to the whole surface area of the colon. c-Met is a human CM tyrosine kinase that is overexpressed in the early stages of colorectal adenoma-carcinoma progression. GE-137 is a fluorescently labeled peptide agent with a high affinity for c-Met. After being conjugated to a fluorescent cyanine dye and administered to mice and human patients, fluorescence colonoscopy enabled visualization of neoplastic polyps [214]. The altered pH_e of cancer tissue could lead to drug resistance and is considered an imaging target. One study used fluorescent probes, two-photon probes (XBH1–3), and a two-photon microscope for the in situ measurement of pH_e. Ex vivo and in vivo results suggested that the XBH1 platform selectively stained cells in the acidified cancer tissue. This probe could directly monitor pH values both inside and outside cells in the colon cancer tissue as well as provide information on morphological aspects [215].

6.4 Targeted prostate cancer imaging

Prostate cancer (PCa) is the second-most frequently diagnosed solid-organ malignancy in men in the United States and the second-most common in males worldwide. Age range (50–74 years), race (African American race), and family history (e.g., BRCA mutations) are the most established risk factors for prostate cancer [216,217]. FDA-approved prostate-specific antigen (PSA or human kallikrein-3), ProPSA, and prostate cancer antigen 3 are noninvasive biomarkers that are currently used for prostate cancer detection [218]. However, most modalities have poor sensitivity and specificity at low PSA levels. Advancements in the field of molecular imaging are important for developing multimodality imaging for biopsy guidance aimed at early detection of PCa, or recurrence post treatment (Table 4.3).

PSMA is a membrane glycoprotein that is strongly upregulated at all stages of PCa. Numerous studies have employed Abs targeting PSMA for improving the imaging sensitivity. At present, only the radiolabeled anti-PSMA Ab targeting the intracellular epitope (7E11) (ProstaScint, Jazz Pharmaceuticals, USA) has been approved by the FDA. The rapid clearance of PET tracer labeled anti-PSMA Ab from off-target tissues made it an ideal tracer for PCa detection, staging, and clinical decisions [219]. The heterogeneity of PCa motivated an increased focus on the tumor vasculature for imaging. Agemy et al. [220] designed a PCa vasculature homing–based (synoptic) targeting agent using iron oxide NPs coated with CREKA, a blood clotting peptide that recognizes the fibrin–fibronectin complexes. The CREKA-PEG-NPs self-amplified their tumor accumulation, enhanced tumor imaging, and allowed for optimized treatment.

Hepsin (HPN) is a type II transmembrane serine protease expressed in the precursor lesions of prostate cancer, high-grade prostatic intraepithelial neoplasia, and hormone-refractory metastatic tumors. HPN binding peptides conjugated to imaging nanoprobes bound to PCa with high affinity in vivo. In situ histochemical analysis of patient tissues demonstrated the potential of this nanoprobe as an imaging agent for PCa [221]. Likewise,

GRPRs are overexpressed in prostate tumor cells. One study conjugated a GRPR bombesin (Bom) peptide to the PET isotope ^{64}Cu. In vitro micro-PET/CT imaging results confirmed the binding specificity of this platform to GRPR on the prostate cancer cell surface. Furthermore, in vivo results demonstrated that these NPs exhibited no acute toxicity in treated mice, suggesting that Bom-PEG-[^{64}Cu] CuS NPs were ideally suited for PET imaging of orthotopic prostate tumors [222]. Another study synthesized SP204 and PC204 peptide-conjugated SPIONs that accumulated in a PCa xenograft model, with potential for PCa-targeted imaging and diagnosis [223].

The robust molecular structure of the tobacco mosaic virus (TMV) offers a versatile platform for theranostic applications. The Hu group [224] synthesized a bimodal imaging agent by loading the internal cavity of TMV self-assembled NPs with an NIRF dye Cy7.5 dysprosium ions (Dy$^{3+)}$ to produce a complex. The imaging probe was then conjugated with Asp-Gly-Glu-Ala (DGEA) peptide that targets integrin α2β1. NIRF imaging and T$_2$-mapping (using ultra-high-field MRI [UHFMRI]) confirmed that this biocompatible probe effectively targeted PC-3 pancreatic cancer (PC) cells and tumors. The Dy-Cy7.5-TMV-DGEA was suitable for multiscale MRI scanning of the entire body, particularly in the context of UHFMRI.

6.5 Targeted pancreatic cancer imaging

PC is a gastrointestinal tumor and is the fourth-leading cause of cancer mortality in the United States due to its late diagnosis, early metastasis, and resistance to chemotherapy. The 5-year patient survival rate for all patients is less than 5% [225]. Based on NIH statistics, 55,440 newly diagnosed cases and 44,330 deaths were reported in the year 2018. The pancreas acts as both an endocrine and exocrine gland. Tumors originating from endocrine tissue are termed islet cell tumors (or neuroendocrine). More than 90% of PC tumors originate from the ductal epithelium of the pancreas, called pancreatic ductal adenocarcinoma (PDAC) [226,227].

Several tyrosine kinases, including VEGFR-2, c-KIT, FGFR-1, colony-stimulating factor 1 receptor (CSF1R), and SRC, are overexpressed on the surface of PC cells. Other receptors (such as SRC, CSF1R, VEGFR-2, c-KIT, PDGFR, TβRI, TβRII, and FGFR-1) are also overexpressed on the surface of PC cells [228,229]. PC is diagnosed in three stages (I, II, and III). The relative expression of targetable molecules differs in each stage [230] (Table 4.3). PC has a dense tumor-stromal barrier, which limits diffusion and the accessibility of contrast agents (as well as drugs) to the cancer cells. Thus, targeting both stromal and PC cells is required. In PC, insulin-like growth factor 1 receptor (IGF1R) is overexpressed in both stromal and tumor cells. Zhou et al. [231] used iron oxide NPs, IGF1 as a ligand, and doxorubicin to form IGF1-IONPx-Dox. Noninvasive MRI results demonstrated that it could act as an effective theranostic system to improve PC targeted imaging and therapy.

P32 (or gC1qR) is a multifunctional cellular receptor protein overexpressed on the surface of PC. The Jiang group designed multifunctional core—shell magnetic nanospheres prepared from iron oxide NPs and silica labeled with FITC and LyP-1 peptide that targets the P32 receptor called $(Fe_3O_4@SiO_2\text{-}FITC@mSiO_2\text{-}LyP\text{-}1)$ (Fig. 4.15). In vivo MRI and fluorescence imaging confirmed specific accumulation of the designed nanospheres in the tumor tissue, allowing MR imaging of orthotopic PC xenografts [232]. Despite its name, prostate stem cell antigen (PSCA) has been reported to be overexpressed in primary pancreatic ductal adenocarcinoma. PSCA has been employed to distinguish PC from chronic pancreatitis, and higher PSCA levels have been correlated with poor prognosis and metastasis of PCa. Zettlitz et al. [233] developed a dual-labeled probe based on anti-PSCA A2 cys-diabody (A2cDb) with a specific conjugation site for IRDye800CW and random [124]I-labeling ([124]I-A2cDb-800). In mice bearing PC xenograft tumors, immunoPET allowed noninvasive, whole-body imaging to localize PCs, and NIRF image guidance could allow identification of tumor margins during resection.

Another overexpressed marker in PCa is the receptor for advanced glycation end products (RAGE), which plays a critical role in the transition of premalignant epithelial precursor cells to pancreatic ductal adenocarcinoma. The Kim group has synthesized a fluorescent dye (Cy5), labeled anti-RAGE scFv antibody, with a high binding affinity to murine RAGE and no internalization in PC cell lines. The anti-RAGE scFv successfully visualized RAGE expression in a $KRAS^{G12D}$ mouse bearing PC tumors. In vivo biodistribution studies used the ^{64}Cu-labeled scFv Ab fragment in a syngeneic mouse model, demonstrating receptor-specific uptake in RAGE-overexpressing tumors. PET imaging data showed anti-RAGE scFv had a high affinity to RAGE in vivo [234].

6.6 Targeted bladder cancer imaging

Nearly 380,000 new cases and 150,000 deaths attributed to bladder cancer are reported annually. Bladder cancer is the fifth most common type of cancer and is responsible for nearly 3% of all cancer-related deaths in the

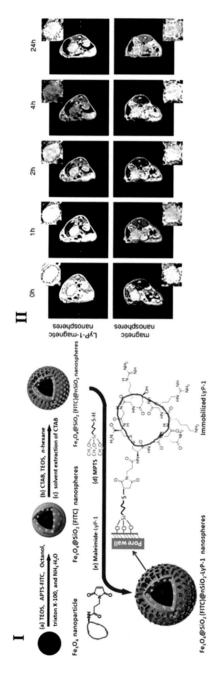

Figure 4.15 *Tumor targeting with multifunctional silica nanospheres.* (I) Synthetic route and structure of multifunctional nanospheres. (A) Coating a layer of FITC-incorporated silica via the co-condensation of TEOS and APTS-FITC. (B) Further growth of a CTAB/SiO2 composite layer using CTAB as a structure-directing agent. (C) Removal of CTAB producing mesopores in the outer shell. (D) Insert thiol groups via the surface modification of NS with MTPS. (E) Immobilization of LyP-1 via the "Click" reaction between thiol groups anchored on the NS and the terminal maleimide group in the cyclic LyP-1 derivative. (II) T2 weighted MRI of orthotopic pancreatic cancer before and after adminis-tration of the Fe₃O₄@SiO₂-FITC@mSiO₂ or Fe₃O₄@SiO₂-FITC@mSiO₂-LyP-1 systemically at different time points (The inset is enlarged picture of corresponding tumor region). *(Reproduced with permission Jiang Y, et al. Magnetic mesoporous nanospheres anchored with LyP-1 as an efficient pancreatic cancer probe. Biomaterials 2017;115:9—18. Copyright 2017, Elsevier).*

United States [235]. Bladder cancer develops in two distinct forms, papillary and nonpapillary, which are pathologically and clinically distinct. The majority of bladder cancers are superficial papillary lesions (NMIBC: non-muscle-invasive bladder cancer) that originate from hyperplastic changes in the mucosa (referred to as low-grade intraepithelial neoplasia) and account for approximately 70%—80% of cases. In the early stages of NMIBC, tumors penetrate the epithelial basement membrane but have not invaded the bladder wall. The opposite is true for most high-grade, muscle-invasive bladder cancers. These tumors can be multifocal and tend to recur after local excision. However, they usually do not metastasize to other organs [236].

Aggressive bladder cancers are usually solid nonpapillary types that originate from in situ precursor lesions (i.e., dysplasia or severe carcinoma in situ). These tumors frequently give rise to distant organ metastasis and are more likely to invade the bladder wall. Clinically, the nonpapillary and papillary forms are separately classified; however, they do have some overlap. Patients with external papillary tumors generally experience multiple recurrences, but only a small fraction progress to high-grade invasive bladder tumors. Conversely, the majority of high-grade invasive bladder cancers develop in patients with no history of superficial papillary lesions. This dual-track concept of bladder carcinogenesis was developed based on correlations between pathological and clinical observations [236,237].

Classification of bladder cancer into different subtypes is based on several factors. Damrauer et al. divided bladder cancer into luminal and basal subtypes [238], while Sjödahl et al. [239] classified bladder cancer, according to four mRNA expression profiles, into five major subtypes: urobasal A (UroA); urobasal (UroB); genetically unstable; squamous cell carcinoma-like; and infiltrating. According to the Cancer Genome Atlas (TCGA), bladder cancer can contain four defined expression clusters (I—IV). Choi et al. [240] classified bladder cancer into three luminal, basal, and p53like subtypes. Fig. 4.16 schematically depicts the subtype classification, overlap between them, subtype markers, and possible targets.

Studies have shown that CD47 is overexpressed in more than 80% of bladder cancer cells. CD47 binds to the signal regulatory protein α, which is expressed on dendritic cells and macrophages to provide signals to prevent phagocytosis. Targeting CD47 in the cancer tissue can be accomplished using different ligands (such as anti-CD47 Abs). To evaluate the expression and function of CD47 in bladder cancer, Pan et al. evaluated fluorescently

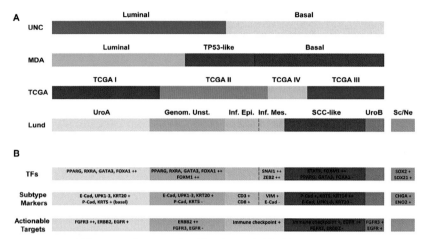

Figure 4.16 *Different subtypes classification of bladder cancer. Genom. Unst.,* genomically unstable; *Inf. Epi.*, infiltrated epithelial; *Inf. Mes.*, infiltrated mesenchymal; *MDA*, MD Anderson Cancer Center; *Sc/Ne*, small cell/neuroendocrine; *SCC*, squamous cell carcinoma; *TCGA*, the Cancer Genome Atlas; *TFs*, transcription factors; *UNC*, University of North Carolina; *UroA*, urobasal A; *UroB*, urobasal B. *(Adapted with permission Aine M, et al. On molecular classification of bladder cancer: out of one, many. Eur Urol 2015;68(6):921—923. Copyright 2015, European Urology).*

labeled anti-CD47 Ab as an intravesical imaging contrast agent. The results of fluorescence imaging, confocal microscopy, and cystoscopy showed the imaging agent possessed high sensitivity and specificity for CD47-targeted imaging [242].

Chemokine receptors (CKRs) are a superfamily of small transmembrane G-protein coupled receptors involved in inflammatory and immune reactions. Different chemokine receptors, including CCR1-CCR10, CXCR1-CXCR6, XCR1, and CX3CR1, have been identified. CXCR4 is the only type of CKR upregulated in MIBC tissue samples. Currently, CXCR4 could be a new molecular probe target with a high affinity for imaging of high-grade superficial bladder cancer. Nishizawa et al. [243] used T140 (14-mer peptide) as an antagonistic ligand for developing a TY14003 molecular probe for targeting CXCR4. The in vivo results of fluorescent imaging indicated that the probe was promising for the detection of flat high-grade superficial bladder cancer lesions.

NMIBC lesions are generally localized to the bladder lumen, while a targeted imaging probe can only detect luminal surface biomarkers. CA9 and CD47 are biomarkers expressed on the luminal surface. Davis et al. synthesized gold-silica NPs as surface-enhanced Raman-scattering

(SERS)-capable NPs targeted with Abs s420-anti-CA9, s440-anti-CD47, and s421-anti-IgG4 for active and passive targeting (Fig. 4.17). The main results of this study were (1) evidence of passive targeting of intravesical NPs; (2) the EPR effect operated for topically applied NPs; and (3) the bladder tissue could be classified as normal or cancerous using multiplexed molecular SERS imaging [244]. More examples of targeted imaging systems for the detection of bladder cancer are summarized in Table 4.3.

6.7 Targeted brain cancer imaging

Brain cancer describes a heterogeneous group of primary and metastatic tumors occurring in the central nervous system. The annual incidence of primary malignant brain tumors is approximately 24,000 cases worldwide. Brain cancer is the leading cause of death in children under the age of 15 [245]. The failure of early diagnosis is mostly due to the absence of targeted imaging systems with high selectivity, and poor treatment outcome is due to the failure of current chemotherapy regimens or incomplete surgical resection (because of the inherent infiltrative character of brain tumors) [246]. Primary brain tumors (composed of cells derived from astrocytes, oligodendrocytes, or ependymal cells) are known as astrocytoma, oligo-dendrogliomas, and ependymomas, respectively [247].

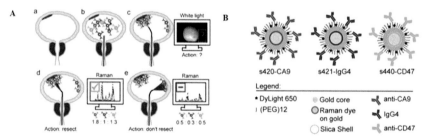

Figure 4.17 *SERS NPs for bladder cancer imaging.* (A) Proposed application of intra-luminal SERS NPs. (a) Patient presents with potential NMIBC (red color tissue). (b) Before cystoscopy, intraluminal SERS NPs are administered. Each NP color represents a different targeting mechanism (passive, blue; CA9, red; and CD47, green). (c) Patient receives standard of care guided transurethral resection. Regions ambiguous on white light cystoscopy (WLC) are subsequently interrogated with Raman endoscopy. (d, e) Based on absolute and relative binding levels of each channel, flat lesions can be identified, and cancer tissue is resected. (B) Schematic representation of the SERS NPs. The blue IgG4 NPs are used as a negative experimental control for the active binding of CA9-and CD47-targeted SERS NPs. (*Adapted with permission Davis RM, et al. Surface-enhanced Raman scattering nanoparticles for multiplexed imaging of bladder cancer tissue permeability and molecular phenotype. ACS Nano 2018;12(10):9669—9679. Copyright 2018, American chemical society*).

Unlike the normal capillaries in the brain, the tight junctions between the ECs of brain tumors are seriously compromised, producing a leaky tumor vasculature, while the high intratumoral interstitial pressure limits drug penetration from the bloodstream into the brain tumor. Moreover, the remaining blood—brain barrier (BBB) limits the transportation of targeted agents. To overcome these problems, multi-targeting imaging probes must display high permeability across the BBB and overcome other penetration impediments. Targeting brain tumors can be improved by targeting receptors, such as integrin $\alpha V\beta 3$, or aminopeptidase N. These receptors are distributed on proliferating ECs within the brain tumor (sites in direct contact with circulating NPs in the bloodstream) and are overexpressed on brain capillary ECs [100,248,249].

Glioblastoma multiforme (GBM) is the most malignant grade (IV) of astrocytoma and requires complete surgical resection for long-term cures. Although traditional chemotherapy does not work well against GBM, innovative dual-targeted imaging and therapy nanoconstruct have been investigated with high permeability across the BBB. Ni et al. [250] developed a bimodal imaging agent for MR/fluorescence imaging of intracranial GBM, benefiting from the MRI and upconversion luminescence (UCL) capabilities of upconversion nanoparticles (UCNPs). CD13 is overexpressed in glioma and can be recognized by a tumor-homing NGR peptide motif. Huang et al. [251] synthesized an ANG-conjugated PEG-CdSe/ZnS quantum dot-based imaging probe. Fluorescence imaging results showed that the probe could cross the BBB and target CD13-overexpressing glioma tumors. The PEGylated UCNPs were modified with angiopep-2 (ANG/PEG-UCNPs) as a targeting ligand with high affinity to the low-density lipoprotein receptor-related protein, overexpressed on both BBB and glioblastoma cells. The ANG/PEG-UCNP platform displayed higher transcytosis across the BBB and endocytosis into glioblastoma cells compared with nontargeted PEG-UCNPs, with no significant cytotoxic effect. The MR images of glioblastoma-bearing mice showed the T_1-weighted contrast was enhanced at the tumor site 1 h post injection of ANG/PEG-UCNPs. The tumor was barely visible in mice injected with PEG-UCNPs or the clinically employed Gd-DTPA contrast agent. The results were better than those with the commonly used fluorescent dye 5-ALA. In a similar study taking advantage of the EPR effect in brain tumors and its angiogenic blood vessels, Li et al. [252] fabricated a dual-modality Gd—Ag_2S nanoprobe to take advantage of the deep tissue penetration of MR and high spatiotemporal resolution of fluorescence imaging, to help surgeons conduct more precise surgery for GBM.

However, these studies did not address how the NPs could cross the tight junctions of the BBB. Diaz et al. [253] employed MRI-guided transcranial focused ultrasound (TcMRgFUS) as a noninvasive technique to increase the permeability of the BBB to allow SERS imaging. They used silica shell-coated GNPs, where after BBB disruption using TcMRgFUS, the delivery of 50 and 120 nm GNPs to the tumor periphery was achieved without any vascular damage. This approach could pave the way for the specific delivery of a wide range of therapeutic and diagnostic agents. Applying a resonating magnetic field allowed magneto–responsive nano-platforms, such as magnetic-fluid-loaded liposomes (MFLs), to target and monitor malignant brain tumors [254]. The in vivo results showed that after 4 h exposure to a focused $190\ Tm^{-1}$ magnetic field gradient, MFLs could pass through the BBB and accumulated only in U87 human glioblastoma xenografts, and were retained therein for almost 24 h, as shown by MRI. There was no sign of MFLs in other areas of the brain (Fig. 4.18).

GLUT-1 and ASCT2 (an important L-isomer-selective amino acid transporter) are found in high density in the BBB and in brain tumors. Zhang et al. [255] prepared CDs tagged with L-Asp, glucose, and/or L-Glu. These groups allowed the CDs to cross the BBB via ACT2 and GLUT-1 transporters. Because the RGD tripeptide is known to act as an αVβ3 integrin targeting agent, they asked whether CD-Asp, could also act as an RGD-like functional group and bind to the αVβ3 integrin on the immature ECs in the glioma. In vitro and in vivo results showed that the CD-Asp NPs could act as an excellent fluorescence imaging and targeting agent for safe and noninvasive glioma imaging.

SPIONs possess negative-contrast capability in MRI and have emerged as a versatile agent in magnetic targeting. Xu et al. [256] described a theranostic liposome (QSC-Lip) preparation based on QDs, SPIONs, and cilengitide (a cyclic RGD pentapeptide) for in vivo dual-MRI/NIR imaging. The data revealed that the QSC-Lip imaging probe not only produced a negative-contrast enhancement in gliomas using MRI but also created tumor-localized fluorescence under magnetic targeting and could be used to guide the surgical resection of the glioma. More examples of targeted brain cancer imaging are summarized in Table 4.3.

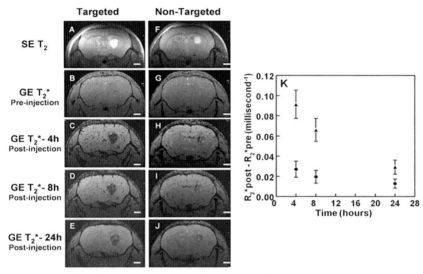

Figure 4.18 *Glioblastoma imaging with magnetic field responsive MFLs and MRI.* Series of brain MR images versus time from glioblastoma-bearing mice injected with MFLs and at 4 h post injection receiving magnetic targeting by external application of 0.4-T magnet (A–E) or not (F–J); the spin-echo T_2–weighted (SE T_2) baseline acquisitions performed before MFLs injection show the tumor locations as hyperintense lesions (A,F); the T_2*-weighted gradient echo (GE T_2*) sequences reveal the presence of the contrast agent as hypointense areas (B–E, G–J); the persistence of the hyposignal at the targeted tumor level remains clearly visible 24 h post injection (E). Relaxation rate difference (R_2*$_{post}$ − R_2*$_{pre}$) for the targeted (▲) and nontargeted (●) tumors as a function of the time period following MFLs administration (K); the magnet was removed at 4 h for the targeted tumor. White bars represent 1 mm. *(Reproduced with permission Marie H, et al. Superparamagnetic liposomes for MRI monitoring and external magnetic field-induced selective targeting of malignant brain tumors. Adv Func Mater 2015;25(8):1258–69. Copyright 2015, Wiley).*

6.8 Targeted ovarian cancer imaging

Ovarian cancer is known as the "silent lady killer" and is the fifth-leading cause of cancer-related deaths in women. Because this cancer is frequently diagnosed during later stages of the disease (stage 3 or 4), it has the highest morbidity and mortality of all gynecological cancers. Epithelial ovarian cancer (accounting for ∼90% of cases) is classified into four histological subtypes: serous, endometrioid, clear-cell, and mucinous carcinomas. Of these types, high-grade serous carcinoma is the most commonly diagnosed, and unlike the other subtypes, it probably originates in the fallopian tubes [257]. The risk of developing ovarian cancer is determined by genetic factors, age, postmenopausal hormonal therapy, infertility, and nulliparity.

In terms of screening, germline mutations in BRCA1 or BRCA2 present a high risk for developing ovarian cancer. In women with an average risk of developing ovarian cancer, the biomarker CA125 has been the primary focus for screening. The combination of CA125 blood test and radiographic imaging (transvaginal ultrasonography) has been evaluated as a screening strategy [257].

Based on grade, size, symptoms, etc., several markers including VEGFR, EGFR, PDGF, and their receptors (PDGFR, KIT pathways, ERBB2, and α-folate receptor [αFR]) have been selected to be implemented in targeted therapy [258]. These therapeutic strategies could be improved by using targeted imaging techniques. SPECT and PET are molecular imaging techniques that have been used in the imaging of ovarian cancer. Ovarian cancer-specific molecules including cell surface receptors, hormone receptors, receptor tyrosine kinases, angiogenic and immune-related factors can be labeled using radioactive nuclides [258]. Although ^{18}F-FDG PET has been studied in the diagnosis of ovarian cancer, it is not thought to be a good option for the primary diagnosis of ovarian cancer. Hence, discovering other options is necessary. Some ovarian tumors show overexpression of ERα ($\sim$70% of patients). The use of PET imaging mediated by ^{18}F-FES demonstrated that this platform could provide reliable information about tumor ERα status and whether endocrine therapy could be employed [259]. HER3 overexpression has been found to be a mediator of tumor resistance to HER1 and HER2-targeted therapies in both breast and ovarian cancer. However, imaging of this receptor using a radiolabeled anti-HER3 mAb showed a long biological half-life and relatively poor tumor penetration. The Chiara Da Pieve group chose to use an affibody with rapid clearance by the kidneys, biocompatibility, and good specificity and affinity. They developed an [^{18}F] aluminum fluoride radiolabeling procedure for the HER3-targeted affibody ($Z_{HER3:8698}$). This platform showed successful tumor targeting with clear visualization of HER3-overexpressing xenografts in tumor-bearing mice, 1 h post injection [260]. The overexpression of folate receptor-α (FR-α) found in 90%$-$95% of epithelial ovarian cancers prompted the investigation of an FR-α-targeted fluorescent agent for intraoperative tumor-specific fluorescence imaging in ovarian cancer surgery. Nanoemulsions (NEs) were loaded with imaging contrast and decorated with folate-PEG$_{3400}$-DSPE in platinum (Pt) resistant ovarian cancer cells [261]. Another study used folate-FITC as an FR-α-targeted fluorescent imaging agent in patients with ovarian cancer [262]. They proved that FR-$\alpha$$-$FITC had a good pharmacodynamic profile after systemic administration in patients and could improve tumor staging and allow real-time visualization of the tumor tissue during surgery.

Combining the overexpressed CA-125 membrane marker with ultrasound (US) contrast agents could allow the detection of early-stage ovarian cancer. The Yong Gao group formulated CA-125-targeted nanobubbles (NBs) to detect CA-125+ ovarian cancer [263]. Their results demonstrated that the targeted NBs were stable, specific, and selectively bound to CA-125+ ovarian cancer cells in vitro, with strong accumulation in ovarian cancer tissue in vivo and long-lasting contrast enhancement. Human epididymis protein 4 (HE4) is one of two US FDA-approved serum biomarkers in ovarian cancer. Tissue concentrations of HE4 are greater than serum concentrations, and hence, HE4 may be a target for ovarian cancer imaging. Recently Williams et al. [264] developed a carbon nanotube (CNT)—based probe using an immobilized Ab that recognized HE4. NIR bandgap photoluminescence from CNTs between 800 and 1600 nm successfully allowed detection of HE4 in patient serum and ascites samples and in orthotopic murine models of ovarian cancer. HER-2 is expressed in a high percentage of ovarian cancers, and systemic delivery of an HER-2 affibody attached to magnetic iron oxide nanoparticles (IONPs) into mice bearing HER-2 positive SKOV3 tumors demonstrated potential for image-guided surgery [265]. However, it would be desirable to simultaneously target multiple cell surface biomarkers to increase the specificity and sensitivity for ovarian cancer detection (Table 4.3).

7. Tumor-specific imaging probes in clinical trials

There are many nanodelivery vehicles in ongoing clinical trials for the delivery of therapeutic agents to tumors. However, most are not typically surface-modified with targeting moieties or equipped for tumor detection and imaging. Tumor-selective imaging probes need to satisfy critical safety and toxicity standards and overcome limitations, such as suboptimum pharmacokinetics, resource-intensive scale-up, reimbursement issues, and an evolving regulatory framework for good manufacturing practices. Addressing these criteria is essential for evaluating the probes that are undergoing preclinical testing or are transitioning into early-phase clinical trials. Some of these probes are being investigated in phase 1 clinical trials in patients with solid tumors, while other specific cancer indications are being explored in advanced clinical trials (phases 2 and 3) [328] (summarized in Table 4.4).

Bevacizumab (Avastin) is an anti-VEGF-A MAb that is used in the clinic for several purposes. PET imaging using (^{89}Zr)-bevacizumab has indicated

Table 4.4 Summary of nanotechnology-assisted cancer-specific molecular imaging in clinical trials.

Imaging modality	Nanoplatform	Cancer type	Receptor	Status	ClinicalTrials.gov identifier
PET/CT	[¹⁸F] Fluoroestradiol (FES)	Breast cancer	ER+	Phase 1	NCT02559544
PET/CT	^{68}Ga–NOTA–BBN–RGD	Breast cancer	GRPR	Phase 1	NCT02749019
PET	[¹⁸F]–ML-10	Metastatic brain cancer	EPR	Phase 2	NCT00805636
PET/CT	(^{99}m Tc) ECDG	Lung cancer	EPR	Phase 3	NCT01394679
PET/CT	^{68}Ga–NOTA–3P–TATE–RGD	Lung cancer	Integrin $\alpha v\beta 3$	Early phase 1	NCT02817945
PET/CT	^{124}I	Breast cancer	NIS	Early phase 1	NCT01360177
PET/CT	^{89}Zr–trastuzumab	Breast cancer	HER2	Phase 1	NCT02286843
PET	^{64}Cu– TP3805	Bladder carcinoma	EPR	Early phase 1	NCT03039413
PET	^{68}Ga–labeled HBED–CC PSMA	Prostate cancer	PSMA	Phase 1 phase 2	NCT02611882
dPET–CT	^{18}F-FMISO	NSCLC	hypoxia	–	NCT01617980
PET	^{89}Zr–trastuzumab	Breast cancer	HER2	–	NCT01081600
PET/CT	(^{18}F-3c) ([¹⁸F]ISO-1)	Breast cancer	Sigma-2 receptor	Phase 1	NCT02762110
PET	^{64}Cu–DOTA–AE105	Breast, prostate and bladder cancer	uPAR	Early phase 1	NCT02139371
PET/CT	[¹⁸F]ISO-1	Breast cancer	Sigma-2 receptor	–	NCT03057743
IOI	OTL38	Ovarian cancer	Folate receptor-α	Phase 2	NCT02317705
PET	^{68}Ga–NOTA–AE105	Breast, prostate and bladder cancer	uPAR	Phase 1	NCT02437539
PET	^{89}Zr–GSK2849330	Cancers	HER3	Phase 1	NCT02345174

Continued

Table 4.4 Summary of nanotechnology-assisted cancer-specific molecular imaging in clinical trials.—cont'd

Imaging modality	Nanoplatform	Cancer type	Receptor	Status	ClinicalTrials.gov identifier
PET	^{89}Zr-AMG211	Gastrointestinal cancer	CEA, CD66e and CD3	Phase 1	NCT02760199
PET/CT	^{68}Ga-PSMA	Prostate cancer	PSMA	Phase 2	NCT03689582
SPECT/CT	^{99}mTc-ABH2	Breast cancer	HER2	Early phase 1	NCT03546478
MRI/PET	^{64}Cu-MM-302	Brain solid tumors	HER2	Early phase 1	NCT02735798
PET/CT	^{18}F-EF5	Ovarian cancer	Hypoxia	—	NCT01881451
PET/CT	^{68}Ga-NODAGA-Ac-cys-ZEGFR:1907	Cancers	EGFR	—	NCT02916329
NIRF	OTL38	Lung cancer	Folate receptor	Phase 2	NCT02872701
PET	^{89}Zr-labeled KN035	Solid tumors	PD-L1	—	NCT03633804
PET	^{68}Ga-labeled F(ab') 2-trastuzumab	Solid tumors	HER2	Phase 1	NCT00613847
PET/CT	^{68}Ga-PSMA	Recurrent prostate carcinoma	PSMA	Phase 3	NCT03582774
PET/CT	^{18}F-FMISO	NSCLC	Hypoxia	—	NCT02016872
PET/CT	^{89}Zr-daratumumab	Multiple Myeloma	CD38	Phase 1 phase 2	NCT03665155
PET	^{18}-F-MISO	Prostate adenocarcinoma	Hypoxia	Phase 2	NCT01898065
PET/CT	^{18}F-DCFPyL	RCC	PSMA	—	NCT02687139
NIRF	Bevacizumab-IRDye800CW	Breast cancer	VEGF	Phase 1	NCT01508572

PET/CT	^{89}Zr-trastuzumab	Breast cancer	HER2	–	NCT02286843
NIRF	Bevacizumab-IRDye800CW	Rectal cancer	VEGF	Phase 1	NCT01972373
NIRF	Indocyanine green	Lung cancer	EPR	Phase 1	NCT00264602
PET/CT	^{18}F-FDG	NSCLC	Metabolism targeting	Phase 3	NCT02938546
PET	^{18}F-Fluoroazomycin arabinoside	Tongue cancer	Hypoxia	Phase 1 phase 2	NCT03181035
Theranostic	^{177}Lu-PP-F11N	Thyroid cancer	Cholecystokinin-2 receptors	Phase 1	NCT02088645
PET/CT	^{18}F-DCFPyL	Prostate cancer	PSMA	Early phase 1	NCT02691169
Ultrasound imaging	Perflutren lipid Microsphere	Prostate cancer	EPR	Phase 2	NCT02967458
PET/CT	^{18}Fluciclatide	Solid tumors	αvβ3	Phase 1 phase 2	NCT01176500
PET/CT	^{64}Cu-plerixafor	Cancers	CXCR4	Early phase 1	NCT02069080
Ultrasound imaging	BR55	Prostate cancer	VEGFR2	Phase 1 phase 2	NCT02142608
PET/MRI	[^{89}Zr]-Df-trastuzumab	Breast cancer	HER2	Early phase 1	NCT03321045
PET/CT	111 in-folic acid	Prostate cancer	Folate receptor	–	NCT00003763
NIRF	OTL38	Ovarian cancer	Folate receptor	Phase 3	NCT03180307

*Abbreviations: [^{18}F]-ML-10, 2-(5-fluoro-pentyl)-2-methyl-malonic-acid; (^{18}F-3c) ([^{18}F]ISO-1), N-(4-(6,7-dimethoxy-3,-4-dihydroisoquinolin-2(1H)-yl) butyl)-2-(2-[18F]-fluoroethoxy)-5-methylbenzamide; ECDG, ethylenedicysteine-deoxyglucose; GRPR, gastrin-releasing peptide receptor; NIRF, near infrared fluorescent image NIS, [Na + I- symporter, sodium iodide symporter]; NSCLC, non-small-cell-lung cancer; PSMA, prostate specific membrane antigen; RCC, renal cell carcinoma; uPAR, urokinase plasminogen activator receptor. All information obtained from (https://clinicaltrials.gov).

that VEGF-A is a suitable target for imaging purposes in various tumor types. For the first time, Weele et al. [329] developed and tested the safety of clinical-grade fluorescent-labeled Bevacizumab-800CW for noninvasive NIFR imaging of VEGF-A in patients with high-grade dysplasia in Barrett's esophagus. The aim of this project was to validate the formulation, production, quality control, stability, extended characterization, and preclinical safety of a fluorescent imaging agent suitable for first-in-human application (Clinical Trial identifier: NCT02129933).

C dots (Cornell dots) are 6–7 nm diameter, core–shell, hybrid silica particles that could allow simultaneous PET/optical imaging for the detection of metastatic melanoma. One study reported an ultrasmall, cancer-selective, silica-based imaging probe, which was recently approved for the first-in-human clinical trials and could overcome several limitations of conventional imaging probes. This multimodal platform consisted of Cy5 fluorescent dye and ^{124}I in a nanostructure functionalized with the cRGDyK peptide that targets $\alpha v\beta 3$ integrin receptors. As part of a larger pilot study consisting of 30 patients from Memorial Sloan Kettering Cancer Center, these probes were evaluated for intraoperative mapping of sentinel lymph nodes in patients with melanoma, breast, cervical, and uterine cancer. Its applications included real-time lymphatic drainage patterns and intraoperative detection and imaging of nodal metastases of melanoma (Clinical Trial identifier: NCT02106598) [328,330].

The kinase insert domain receptor (KDR) is an important regulator of neoangiogenesis in human tumors. Willmann et al. [331] carried out the first human clinical trial using USMI in 24 women with ovarian cancer and 21 women with breast cancer using KDR targeted contrast microbubbles (MB_{KDR}). The imaging probe was injected intravenously, and USMI was conducted from 5 to 29 min post injection. USMI mediated by MB_{KDR} was well tolerated by all the patients without any safety concerns. Among the 40 patients undergoing analysis, KDR expression determined by immunohistochemical (IHC) staining matched well with the imaging data for both cancer types (EudraCT Number: 2012-000699-40).

References

[1] Nabil G, et al. Nanoengineered delivery systems for cancer imaging and therapy: recent advances, future directions and patent evaluation. Drug Discov Today 2018;24(2):462–91.
[2] Rosenblum D, et al. Progress and challenges towards targeted delivery of cancer therapeutics. Nat Commun 2018;9(1):1410.

[3] Yousef S, et al. Development of asialoglycoprotein receptor directed nanoparticles for selective delivery of curcumin derivative to hepatocellular carcinoma. Heliyon 2018;4(12):e01071.

[4] Danhier F. To exploit the tumor microenvironment: since the EPR effect fails in the clinic, what is the future of nanomedicine? J Contr Release 2016;244:108−21.

[5] Zheng X, et al. Hypoxia-specific ultrasensitive detection of tumours and cancer cells in vivo. Nat Commun 2015;6:5834.

[6] Guzy RD, et al. Mitochondrial complex III is required for hypoxia-induced ROS production and cellular oxygen sensing. Cell Metabol 2005;1(6):401−8.

[7] Primeau AJ, et al. The distribution of the anticancer drug Doxorubicin in relation to blood vessels in solid tumors. Clin Canc Res 2005;11(24):8782−8.

[8] Chitneni SK, et al. Molecular imaging of hypoxia. J Nucl Med: Off Pub Soc Nucl Med 2011;52(2):165.

[9] Cui L, et al. A new prodrug-derived ratiometric fluorescent probe for hypoxia: high selectivity of nitroreductase and imaging in tumor cell. Org Lett 2011;13(5):928−31.

[10] Liu J, et al. Simultaneous fluorescence sensing of Cys and GSH from different emission channels. J Am Chem Soc 2013;136(2):574−7.

[11] Lee MH, et al. Disulfide-cleavage-triggered chemosensors and their biological applications. Chem Rev 2013;113(7):5071−109.

[12] Harris AL. Hypoxia—a key regulatory factor in tumour growth. Nat Rev Canc 2002;2(1):38.

[13] Luo C-H, et al. Bacteria-mediated hypoxia-specific delivery of nanoparticles for tumors imaging and therapy. Nano Lett 2016;16(6):3493−9.

[14] Wang J, et al. A porous Au@Rh bimetallic core−shell nanostructure as an H_2O_2-driven oxygenerator to alleviate tumor hypoxia for simultaneous bimodal imaging and enhanced photodynamic therapy. Adv Mater 2020;32(22):2001862.

[15] Fan Y, et al. Targeted tumor hypoxia dual-mode CT/MR imaging and enhanced radiation therapy using dendrimer-based nanosensitizers. Adv Funct Mater 2020;30(13):1909285.

[16] Li Y, et al. Dual hypoxia-targeting RNAi nanomedicine for precision cancer therapy. Nano Lett 2020;20(7):4857−63.

[17] Zhou Y, et al. In vivo imaging of hypoxia associated with inflammatory bowel disease by a cytoplasmic protein-powered fluorescence cascade amplifier. Anal Chem 2020;92(8):5787−94.

[18] Kwon Y-D, et al. Novel multifunctional 18F-labelled PET tracer with prostate-specific membrane antigen-targeting and hypoxia-sensitive moieties. Eur J Med Chem 2020;189:112099.

[19] Lv Z, et al. Phosphorescent starburst Pt (II) porphyrins as bifunctional therapeutic agents for tumor hypoxia imaging and photodynamic therapy. ACS Appl Mater Interfaces 2018;10(23):19523−33.

[20] Knox HJ, et al. A bioreducible N-oxide-based probe for photoacoustic imaging of hypoxia. Nat Commun 2017;8(1):1794.

[21] Huo D, et al. Hypoxia-targeting, tumor microenvironment responsive nanocluster bomb for radical-enhanced radiotherapy. ACS Nano 2017;11(10):10159−74.

[22] Zhong D, et al. Photosynthetic biohybrid nanoswimmers system to alleviate tumor hypoxia for FL/PA/MR imaging-guided enhanced radio-photodynamic synergetic therapy. Adv Funct Mater 2020;30(17):1910395.

[23] Gatenby RA, Gillies RJ. Why do cancers have high aerobic glycolysis? Nat Rev Canc 2004;4(11):891.

[24] Estrella V, et al. Acidity generated by the tumor microenvironment drives local invasion. Canc Res 2013;73(5):1524−35.

[25] Manchun S, Dass CR, Sriamornsak P. Targeted therapy for cancer using pH-responsive nanocarrier systems. Life Sci 2012;90(11−12):381−7.

[26] Liu J, et al. CO_2 gas induced drug release from pH-sensitive liposome to circumvent doxorubicin resistant cells. Chem Commun 2012;48(40):4869−71.

[27] Gilson RC, et al. Protonation and trapping of a small pH-sensitive near-infrared fluorescent molecule in the acidic tumor environment delineate diverse tumors in vivo. Mol Pharm 2015;12(12):4237−46.

[28] Zhao X, et al. Dual-stimuli responsive and reversibly activatable theranostic nanoprobe for precision tumor-targeting and fluorescence-guided photothermal therapy. Nat Commun 2017;8:14998.

[29] Urano Y, et al. Selective molecular imaging of viable cancer cells with pH-activatable fluorescence probes. Nat Med 2009;15(1):104.

[30] Wang Y, et al. A nanoparticle-based strategy for the imaging of a broad range of tumours by nonlinear amplification of microenvironment signals. Nat Mater 2014;13(2):204.

[31] Mi P, et al. A pH-activatable nanoparticle with signal-amplification capabilities for non-invasive imaging of tumour malignancy. Nat Nanotechnol 2016;11(8):724−30.

[32] Li Z, et al. Near-infrared/pH dual-responsive nanocomplexes for targeted imaging and chemo/gene/photothermal tri-therapies of non-small cell lung cancer. Acta Biomaterialia 2020;107:242−59.

[33] Feng Y, et al. Optical imaging and pH-awakening therapy of deep tissue cancer based on specific upconversion nanophotosensitizers. Biomaterials 2020;230:119637.

[34] Zhao Y, et al. Polystyrene@ poly (ar-vinylbenzyl) trimethylammonium-co-acrylic acid core/shell pH-responsive nanoparticles for active targeting and imaging of cancer cell based on aggregation induced emission. Microchimica Acta 2020;187(3):1−10.

[35] Hsu BYW, et al. pH-activatable MnO-based fluorescence and magnetic resonance bimodal nanoprobe for cancer imaging. Adv Healthc Mater 2016;5(6):721−9.

[36] Huang G, et al. PET imaging of occult tumours by temporal integration of tumour-acidosis signals from pH-sensitive 64 Cu-labelled polymers. Nat Biomed Eng 2020;4(3):314−24.

[37] Nagase H, Visse R, Murphy G. Structure and function of matrix metalloproteinases and TIMPs. Cardiovasc Res 2006;69(3):562−73.

[38] Egeblad M, Werb Z. New functions for the matrix metalloproteinases in cancer progression. Nat Rev Canc 2002;2(3):161.

[39] Choi JW, et al. Matrix metalloproteinase 3 is a stromal marker for chicken ovarian cancer. Oncol Lett 2011;2(6):1047−51.

[40] Schmalfeldt B, et al. Increased expression of matrix metalloproteinases (MMP)-2, MMP-9, and the urokinase-type plasminogen activator is associated with progression from benign to advanced ovarian cancer. Clin Canc Res 2001;7(8):2396−404.

[41] Sternlicht MD, Werb Z. How matrix metalloproteinases regulate cell behavior. Annu Rev Cell Dev Biol 2001;17(1):463−516.

[42] Tu Y, Zhu L. Matrix metalloproteinase-sensitive nanocarriers. In: Smart pharmaceutical nanocarriers. World Scientific; 2016. p. 83−116.

[43] Turk BE, et al. Determination of protease cleavage site motifs using mixture-based oriented peptide libraries. Nat Biotechnol 2001;19(7):661.

[44] Kondo N, et al. Development of PEGylated peptide probes conjugated with 18F-labeled BODIPY for PET/optical imaging of MT1-MMP activity. J Contr Release 2015;220:476−83.

[45] Wang K-H, et al. Optical imaging of ovarian cancer using a matrix metalloproteinase-3-sensitive near-infrared fluorescent probe. PLoS One 2018;13(2):e0192047.

[46] Atukorale PU, et al. Vascular targeting of nanoparticles for molecular imaging of diseased endothelium. Adv Drug Deliv Rev 2017;113:141−56.

[47] Petrini I, et al. ED-B fibronectin expression is a marker of epithelial-mesenchymal transition in translational oncology. Oncotarget 2017;8(3):4914.

[48] Sceneay J, Smyth MJ, Möller A. The pre-metastatic niche: finding common ground. Canc Metastasis Rev 2013;32(3—4):449—64.

[49] Zhou F, et al. Nuclear receptor NR4A1 promotes breast cancer invasion and metastasis by activating TGF-β signalling. Nat Commun 2014;5:3388.

[50] Bae YK, et al. Fibronectin expression in carcinoma cells correlates with tumor aggressiveness and poor clinical outcome in patients with invasive breast cancer. Hum Pathol 2013;44(10):2028—37.

[51] Arnold SA, et al. Urinary oncofetal ED-A fibronectin correlates with poor prognosis in patients with bladder cancer. Clin Exp Metastasis 2016;33(1):29—44.

[52] Malik G, et al. Plasma fibronectin promotes lung metastasis by contributions to fibrin clots and tumor cell invasion. Cancer research; 2010. p. 0008—5472. CAN-09-3312.

[53] Han Z, Lu Z-R. Targeting fibronectin for cancer imaging and therapy. J Mater Chem B 2017;5(4):639—54.

[54] Rybak J-N, et al. The extra-domain a of fibronectin is a vascular marker of solid tumors and metastases. Canc Res 2007;67(22):10948—57.

[55] Zhou Z, et al. MRI detection of breast cancer micrometastases with a fibronectin-targeting contrast agent. Nat Commun 2015;6:7984.

[56] Han Z, et al. Targeted gadofullerene for sensitive magnetic resonance imaging and risk-stratification of breast cancer. Nat Commun 2017;8(1):692.

[57] Abou-Elkacem L, et al. Ultrasound molecular imaging of the breast cancer neo-vasculature using engineered fibronectin scaffold ligands: a novel class of targeted contrast ultrasound agent. Theranostics 2016;6(11):1740.

[58] Wang T-Y, et al. Collagen-targeted theranostic nanosponges for delivery of the matrix metalloproteinase 14 inhibitor naphthofluorescein. Chem Mater 2020;32(9):3707—14.

[59] Kasten BB, et al. Targeting MMP-14 for dual PET and fluorescence imaging of glioma in preclinical models. Eur J Nucl Med Mol Imag 2020;47(6):1412—26.

[60] Edgington LE, et al. Noninvasive optical imaging of apoptosis by caspase-targeted activity-based probes. Nat Med 2009;15(8):967.

[61] Shalini S, et al. Old, new and emerging functions of caspases. Cell Death & Different 2015;22(4):526.

[62] van Genderen HO, et al. Extracellular annexin A5: functions of phosphatidylserine-binding and two-dimensional crystallization. Biochim Biophys Acta Mol Cell Res 2008;1783(6):953—63.

[63] Boersma HH, et al. Past, present, and future of annexin A5: from protein discovery to clinical applications. J Nucl Med 2005;46(12):2035—50.

[64] Ye D, et al. Bioorthogonal cyclization-mediated in situ self-assembly of small-molecule probes for imaging caspase activity in vivo. Nat Chem 2014;6(6):519.

[65] Zhang W, et al. Au hollow nanorods-chimeric peptide nanocarrier for NIR-II photothermal therapy and real-time apoptosis imaging for tumor theranostics. Theranostics 2019;9(17):4971.

[66] Lu C, et al. Preliminary biological evaluation of 18F-AlF-NOTA-MAL-Cys-Annexin V as a novel apoptosis imaging agent. Oncotarget 2017;8(31):51086.

[67] Watanabe H, et al. In-vivo visualization of radiation-induced apoptosis using 125I-annexin V. Nucl Med Commun 2006;27(1):81—9.

[68] Lahorte C, et al. Biodistribution and dosimetry study of 123I-rh-annexin V in mice and humans. Nucl Med Commun 2003;24(8):871—80.

[69] Ke S, et al. Imaging taxane-induced tumor apoptosis using PEGylated, 111In-labeled annexin V. J Nucl Med 2004;45(1):108—15.

[70] Hanahan D, Folkman J. Patterns and emerging mechanisms of the angiogenic switch during tumorigenesis. Cell 1996;86(3):353–64.

[71] Hellström M, et al. Lack of pericytes leads to endothelial hyperplasia and abnormal vascular morphogenesis. J Cell Biol 2001;153(3):543–54.

[72] Ruoslahti E. Specialization of tumour vasculature. Nat Rev Canc 2002;2(2):83.

[73] de Bruijn HS, et al. Acute cellular and vascular responses to photodynamic therapy using EGFR-targeted nanobody-photosensitizer conjugates studied with intravital optical imaging and magnetic resonance imaging. Theranostics 2020;10(5):2436.

[74] Paiva I, et al. Synthesis and analysis of 64Cu-labeled GE11-modified polymeric micellar nanoparticles for EGFR-targeted molecular imaging in a colorectal cancer model. Mol Pharm 2020;17(5):1470–81.

[75] Hong H, et al. In vivo targeting and imaging of tumor vasculature with radiolabeled, antibody-conjugated nanographene. ACS Nano 2012;6(3):2361–70.

[76] Gao W, et al. Targeting and destroying tumor vasculature with a near-infrared laser-activated "nanobomb" for efficient tumor ablation. Biomaterials 2017;139:1–11.

[77] Korpanty G, et al. Monitoring response to anticancer therapy by targeting micro-bubbles to tumor vasculature. Clin Canc Res 2007;13(1):323–30.

[78] Kim H, Goh S-H, Choi Y. Quenched cetuximab conjugate for fast fluorescence imaging of EGFR-positive lung cancers. Biomater Sci 2020;9(2):456–62.

[79] Kaplan RN, et al. VEGFR1-positive haematopoietic bone marrow progenitors initiate the pre-metastatic niche. Nature 2005;438(7069):820.

[80] Dvorak HF. Vascular permeability factor/vascular endothelial growth factor: a critical cytokine in tumor angiogenesis and a potential target for diagnosis and therapy. J Clin Oncol 2002;20(21):4368–80.

[81] Hicklin DJ, Ellis LM. Role of the vascular endothelial growth factor pathway in tumor growth and angiogenesis. J Clin Oncol 2005;23(5):1011–27.

[82] Alitalo K, Carmeliet P. Molecular mechanisms of lymphangiogenesis in health and disease. Canc Cell 2002;1(3):219–27.

[83] Ellis LM, Hicklin DJ. VEGF-targeted therapy: mechanisms of anti-tumour activity. Nat Rev Canc 2008;8(8):579.

[84] Senger DR, et al. Vascular permeability factor (VPF, VEGF) in tumor biology. Canc Metastasis Rev 1993;12(3–4):303–24.

[85] Gerber H-P, Ferrara N. Pharmacology and pharmacodynamics of bevacizumab as monotherapy or in combination with cytotoxic therapy in preclinical studies. Canc Res 2005;65(3):671–80.

[86] Nagengast WB, et al. In vivo VEGF imaging with radiolabeled bevacizumab in a human ovarian tumor xenograft. J Nucl Med 2007;48(8):1313.

[87] van Scheltinga AGT, et al. Intraoperative near-infrared fluorescence tumor imaging with vascular endothelial growth factor and human epidermal growth factor receptor 2 targeting antibodies. J Nucl Med 2011;52(11):1778–85.

[88] Abakumov MA, et al. VEGF-targeted magnetic nanoparticles for MRI visualization of brain tumor. Nanomed Nanotechnol Biol Med 2015;11(4):825–33.

[89] Jayson GC, et al. Molecular imaging and biological evaluation of HuMV833 anti-VEGF antibody: implications for trial design of antiangiogenic antibodies. J Natl Cancer Inst 2002;94(19):1484–93.

[90] Wang Q, et al. Construction of a novel bispecific fusion protein to enhance targeting for pancreatic cancer imaging. Biomaterials 2020;255:120161.

[91] Desgrosellier JS, Cheresh DA. Integrins in cancer: biological implications and therapeutic opportunities. Nat Rev Canc 2010;10(1):9.

[92] Raab-Westphal S, Marshall JF, Goodman SL. Integrins as therapeutic targets: successes and cancers. Cancers 2017;9(9):110.

[93] Han J, et al. Reconstructing and deconstructing agonist-induced activation of integrin αIIbβ3. Curr Biol 2006;16(18):1796—806.

[94] Berghoff AS, et al. αvβ3, αvβ5 and αvβ6 integrins in brain metastases of lung cancer. Clin Exp Metastasis 2014;31(7):841—51.

[95] McCabe N, et al. Prostate cancer specific integrin αvβ3 modulates bone metastatic growth and tissue remodeling. Oncogene 2007;26(42):6238.

[96] Heß K, et al. Correlation between the expression of integrins in prostate cancer and clinical outcome in 1284 patients. Ann Diagn Pathol 2014;18(6):343—50.

[97] Diaz LK, et al. β4 integrin subunit gene expression correlates with tumor size and nuclear grade in early breast cancer. Mod Pathol 2005;18(9):1165.

[98] Sloan EK, et al. Tumor-specific expression of αvβ3 integrin promotes spontaneous metastasis of breast cancer to bone. Breast Canc Res 2006;8(2):R20.

[99] Hosotani R, et al. Expression of integrin αvβ3 in pancreatic carcinoma: relation to MMP-2 activation and lymph node metastasis. Pancreas 2002;25(2):e30—5.

[100] Bello L, et al. αvβ3 and αvβ5 integrin expression in glioma periphery. Neurosurgery 2001;49(2):380—90.

[101] Slack-Davis JK, et al. Vascular cell adhesion molecule-1 is a regulator of ovarian cancer peritoneal metastasis. Canc Res 2009;69(4):1469—76.

[102] Landen CN, et al. Tumor-selective response to antibody-mediated targeting of αvβ3 integrin in ovarian cancer. Neoplasia 2008;10(11):1259—67.

[103] Gruber G, et al. Correlation between the tumoral expression of β3-integrin and outcome in cervical cancer patients who had undergone radiotherapy. Br J Canc 2005;92(1):41.

[104] Adachi M, et al. Significance of integrin α5 gene expression as a prognostic factor in node-negative non-small cell lung cancer. Clin Canc Res 2000;6(1):96—101.

[105] Takayama S, et al. The relationship between bone metastasis from human breast cancer and integrin αvβ3 expression. Anticancer Res 2005;25(1A):79—83.

[106] Hieken TJ, et al. Molecular prognostic markers in intermediate-thickness cutaneous malignant melanoma. Canc: Interdiscip Int J Am Canc Soc 1999;85(2):375—82.

[107] Patsenker E, et al. The αvβ6 integrin is a highly specific immunohistochemical marker for cholangiocarcinoma. J Hepatol 2010;52(3):362—9.

[108] Bates RC, et al. Transcriptional activation of integrin β6 during the epithelial-mesenchymal transition defines a novel prognostic indicator of aggressive colon carcinoma. J Clin Invest 2005;115(2):339—47.

[109] Nieberler M, et al. Exploring the role of RGD-recognizing integrins in cancer. Cancers 2017;9(9):116.

[110] Humphries JD, Byron A, Humphries MJ. Integrin ligands at a glance. J Cell Sci 2006;119(19):3901—3.

[111] Tang L, et al. Radiolabeled angiogenesis-targeting croconaine nanoparticles for tri-modality imaging guided photothermal therapy of glioma. ACS Appl Nano Mater 2018;1(4):1741—9.

[112] Tang R, et al. Tunable ultrasmall visible-to-extended near-infrared emitting silver sulfide quantum dots for integrin-targeted cancer imaging. ACS Nano 2015;9(1):220—30.

[113] Fluksman A, et al. Integrin α2β1-targeted self-assembled nanocarriers for tumor bioimaging. ACS Appl Bio Mater 2020;3(9):6059—70.

[114] Zhao M, et al. A novel αvβ3 integrin-targeted NIR-II nanoprobe for multimodal imaging-guided photothermal therapy of tumors in vivo. Nanoscale 2020;12(13):6953—8.

[115] Li H, et al. Synthesis and preclinical evaluation of a 68Ga-radiolabeled peptide targeting very late antigen-3 for PET imaging of pancreatic cancer. Mol Pharm 2020;17(8):3000—8.

[116] Kim H, et al. Mini-platform for off—on near-infrared fluorescence imaging using peptide-targeting ligands. Bioconjugate Chem 2020;31(3):721—8.

[117] Liu Z, et al. Dual integrin and gastrin-releasing peptide receptor targeted tumor imaging using 18F-labeled PEGylated RGD-bombesin heterodimer 18F-FB-PEG3-Glu-RGD-BBN. J Med Chem 2008;52(2):425—32.

[118] Jin Z-H, et al. Development of the fibronectin—mimetic peptide KSSPHSRN (SG) 5RGDSP as a novel radioprobe for molecular imaging of the cancer biomarker α5β1 integrin. Biol Pharm Bull 2015;38(11):1722—31.

[119] Alvarez P, et al. Regulatory systems in bone marrow for hematopoietic stem/progenitor cells mobilization and homing. BioMed Res Int 2013;2013.

[120] Wittchen ES. Endothelial signaling in paracellular and transcellular leukocyte transmigration. Front Biosci: J & Virtual Lib 2009;14:2522.

[121] Costa MF, et al. CCL25 induces α4β7 integrin-dependent migration of IL-17+γδ T lymphocytes during an allergic reaction. Eur J Immunol 2012;42(5):1250—60.

[122] Kon S, Atakilit A, Sheppard D. Short form of α9 promotes α9β1 integrin-dependent cell adhesion by modulating the function of the full-length α9 subunit. Exp Cell Res 2011;317(12):1774—84.

[123] Barthel SR, et al. Differential engagement of modules 1 and 4 of vascular cell adhesion molecule-1 (CD106) by integrins α4β1 (CD49d/29) and αMβ2 (CD11b/18) of eosinophils. J Biol Chem 2006;281(43):32175—87.

[124] Hession C, et al. Cloning of an alternate form of vascular cell adhesion molecule-1 (VCAM1). J Biol Chem 1991;266(11):6682—5.

[125] Barreiro O, et al. Dynamic interaction of VCAM-1 and ICAM-1 with moesin and ezrin in a novel endothelial docking structure for adherent leukocytes. J Cell Biol 2002;157(7):1233—45.

[126] Schlesinger M, Bendas G. Vascular cell adhesion molecule-1 (VCAM-1)—an increasing insight into its role in tumorigenicity and metastasis. Int J Canc 2015;136(11):2504—14.

[127] Chen Q, Zhang XH-F, Massagué J. Macrophage binding to receptor VCAM-1 transmits survival signals in breast cancer cells that invade the lungs. Canc Cell 2011;20(4):538—49.

[128] Zhang X, et al. PET imaging of VCAM-1 expression and monitoring therapy response in tumor with a 68Ga-labeled single chain variable fragment. Mol Pharm 2018;15(2):609—18.

[129] Patel N, et al. Bimodal imaging of inflammation with SPECT/CT and MRI using iodine-125 labeled VCAM-1 targeting microparticle conjugates. Bioconjugate Chem 2015;26(8):1542—9.

[130] Wang L, et al. Imaging of neurite network with an anti-L1CAM aptamer generated by neurite-SELEX. J Am Chem Soc 2018;140(51):18066—73.

[131] Uddin MI, et al. Real-time imaging of VCAM-1 mRNA in TNF-α activated retinal microvascular endothelial cells using antisense hairpin-DNA functionalized gold nanoparticles. Nanomed Nanotechnol Biol Med 2018;14(1):63—71.

[132] Cheng VW, et al. VCAM-1 targeted magnetic resonance imaging enables detection of brain micrometastases from different primary tumours. Clinical Cancer Research; 2018. clincanres. 1889.2018.

[133] Scalici JM, et al. Imaging VCAM-1 as an indicator of treatment efficacy in metastatic ovarian cancer. J Nucl Med: Off Pub Soc Nucl Med 2013;54(11):1883.

[134] Sugyo A, et al. Uptake of 111In-labeled fully human monoclonal antibody TSP-A18 reflects transferrin receptor expression in normal organs and tissues of mice. Oncol Rep 2017;37(3):1529—36.

[135] Zhao T, et al. Transferrin-directed preparation of red-emitting copper nanoclusters for targeted imaging of transferrin receptor over-expressed cancer cells. J Mater Chem B 2015;3(11):2388—94.

[136] Zhao Y, et al. Tumor-targeted and clearable human protein-based MRI nanoprobes. Nano Lett 2017;17(7):4096—100.

[137] Wang K, et al. Self-assembled IR780-loaded transferrin nanoparticles as an imaging, targeting and PDT/PTT agent for cancer therapy. Sci Rep 2016;6:27421.

[138] Qi H, et al. Transferrin-targeted magnetic/fluorescence micelles as a specific bifunctional nanoprobe for imaging liver tumor. Nanoscale Res Lett 2014;9(1):595.

[139] Zhen Z, et al. Ferritin nanocages to encapsulate and deliver photosensitizers for efficient photodynamic therapy against cancer. ACS Nano 2013;7(8):6988—96.

[140] Yang M, et al. Dragon fruit-like biocage as an iron trapping nanoplatform for high efficiency targeted cancer multimodality imaging. Biomaterials 2015;69:30—7.

[141] Zhu M, et al. Indocyanine green-holo-transferrin nanoassemblies for tumor-targeted dual-modal imaging and photothermal therapy of glioma. ACS Appl Mater Interfaces 2017;9(45):39249—58.

[142] Tseng J-C, Peterson JD. Optical imaging of bombesin and transferrin receptor expression are as effective as 18FDG in assessing drug efficacy. AACR; 2017.

[143] Goswami U, et al. Transferrin—copper nanocluster—doxorubicin nanoparticles as targeted theranostic cancer nanodrug. ACS Appl Mater Interfaces 2018;10(4): 3282—94.

[144] Weitman SD, et al. Distribution of the folate receptor GP38 in normal and malignant cell lines and tissues. Canc Res 1992;52(12):3396—401.

[145] Hai X, et al. Folic acid encapsulated graphene quantum dots for ratiometric pH sensing and specific multicolor imaging in living cells. Sensor Actuator B Chem 2018;268:61—9.

[146] Hassan M, et al. Engineering carbon quantum dots for photomediated theranostics. Nano Res 2018;11(1):1—41.

[147] Gao X, et al. Controllable synthesis of a smart multifunctional nanoscale metal—organic framework for magnetic resonance/optical imaging and targeted drug delivery. ACS Appl Mater Interfaces 2017;9(4):3455—62.

[148] Liu Q, et al. Distinguish cancer cells based on targeting turn-on fluorescence imaging by folate functionalized green emitting carbon dots. Biosens Bioelectron 2015;64:119—25.

[149] Liu X, et al. Biodegradable and crosslinkable PPF—PLGA—PEG self-assembled nanoparticles dual-decorated with folic acid ligands and Rhodamine B fluorescent probes for targeted cancer imaging. RSC Adv 2015;5(42):33275—82.

[150] Yin F, et al. Folic acid-conjugated organically modified silica nanoparticles for enhanced targeted delivery in cancer cells and tumor in vivo. J Mater Chem B 2015;3(29):6081—93.

[151] Zhou B, et al. PEGylated polyethylenimine-entrapped gold nanoparticles modified with folic acid for targeted tumor CT imaging. Colloids Surf B Biointerfaces 2016;140:489—96.

[152] Jin X, et al. Folate receptor targeting and cathepsin B-sensitive drug delivery system for selective cancer cell death and imaging. ACS Med Chem Lett 2020;11(8): 1514—20.

[153] Zhang Z, et al. Facile synthesis of folic acid-modified iron oxide nanoparticles for targeted MR imaging in pulmonary tumor xenografts. Mol Imag Biol 2016;18(4):569—78.

[154] Kadian S, et al. Targeted bioimaging and sensing of folate receptor-positive cancer cells using folic acid-conjugated sulfur-doped graphene quantum dots. Microchimica Acta 2020;187(8):1—10.

[155] McCawley LJ, O'Brien P, Hudson LG. Overexpression of the epidermal growth factor receptor contributes to enhanced ligand-mediated motility in keratinocyte cell lines. Endocrinology 1997;138(1):121—7.

[156] Nicholson R, Gee J, Harper M. EGFR and cancer prognosis. Eur J Canc 2001;37:9—15.

[157] Kitai Y, et al. Cell selective targeting of a simian virus 40 virus-like particle conjugated to epidermal growth factor. J Biotechnol 2011;155(2):251—6.

[158] Tjalma JJ, et al. Molecular fluorescence endoscopy targeting vascular endothelial growth factor A for improved colorectal polyp detection. J Nucl Med 2016;57(3):480—5.

[159] Gao M, et al. Targeted imaging of EGFR overexpressed cancer cells by brightly fluorescent nanoparticles conjugated with cetuximab. Nanoscale 2016;8(32): 15027—32.

[160] Hudson SV, et al. Targeted noninvasive imaging of EGFR-expressing orthotopic pancreatic cancer using multispectral optoacoustic tomography. Cancer research; 2014.

[161] Wang Z, et al. Active targeting theranostic iron oxide nanoparticles for MRI and magnetic resonance-guided focused ultrasound ablation of lung cancer. Biomaterials 2017;127:25—35.

[162] Hay N. Reprogramming glucose metabolism in cancer: can it be exploited for cancer therapy? Nat Rev Canc 2016;16(10):635.

[163] Evans NR, et al. PET imaging of the neurovascular interface in cerebrovascular disease. Nat Rev Neurol 2017;13(11):676.

[164] Dreifuss T, et al. Glucose-functionalized gold nanoparticles as a metabolically targeted CT contrast agent for distinguishing tumors from non-malignant metabolically active processes. In: Nanoscale imaging, sensing, and actuation for biomedical applications XIV. International Society for Optics and Photonics; 2017.

[165] Singh S. Glucose decorated gold nanoclusters: a membrane potential independent fluorescence probe for rapid identification of cancer cells expressing Glut receptors. Colloids Surf B Biointerfaces 2017;155:25—34.

[166] Cheng Y, et al. Development of a deep-red fluorescent glucose-conjugated bioprobe for in vivo tumor targeting. Chem Commun 2020;56(7):1070—3.

[167] Zhao W, et al. Glucose ligand modififed thermally activated delayed fluorescence targeted nanoprobe for malignant cells imaging diagnosis. Photodiagnosis Photodyn Ther 2020;30:101744.

[168] Rasouli R, et al. Preparation and evaluation of new LAT1-targeted USPION to improve sensitivity and specificity in metabolic magnetic imaging of breast cancer. Biointerface Res Appl Chem 2021:10248—64.

[169] Mohamed MM, Sloane BF. Cysteine cathepsins: multifunctional enzymes in cancer. Nat Rev Canc 2006;6(10):764.

[170] Olson OC, Joyce JA. Cysteine cathepsin proteases: regulators of cancer progression and therapeutic response. Nat Rev Canc 2015;15(12):712.

[171] Ryu JH, et al. Non-invasive optical imaging of cathepsin B with activatable fluorogenic nanoprobes in various metastatic models. Biomaterials 2014;35(7):2302—11.

[172] Shim MK, et al. Cathepsin B-specific metabolic precursor for in vivo tumor-specific fluorescence imaging. Angew Chem 2016;128(47):14918—23.

[173] Liu J, et al. Multifunctional metal—organic framework nanoprobe for cathepsin B-activated cancer cell imaging and chemo-photodynamic therapy. ACS Appl Mater Interfaces 2017;9(3):2150—8.

[174] Gaikwad HK, et al. Molecular imaging of cancer using X-ray computed tomography with protease targeted iodinated activity-based probes. Nano Lett 2018;18(3): 1582—91.

[175] Tsvirkun D, et al. CT imaging of enzymatic activity in cancer using covalent probes reveal a size-dependent pattern. J Am Chem Soc 2018;140(38):12010−20.

[176] Bertrand N, et al. Cancer nanotechnology: the impact of passive and active targeting in the era of modern cancer biology. Adv Drug Deliv Rev 2014;66:2−25.

[177] Perou CM, et al. Molecular portraits of human breast tumours. Nature 2000;406(6797):747.

[178] Smith RA, et al. Cancer screening in the United States, 2018: a review of current American Cancer Society guidelines and current issues in cancer screening. CA. A Cancer J Clin 2018;68(4):297−316.

[179] Network CGA. Comprehensive molecular portraits of human breast tumours. Nature 2012;490(7418):61.

[180] Wirapati P, et al. Meta-analysis of gene expression profiles in breast cancer: toward a unified understanding of breast cancer subtyping and prognosis signatures. Breast Canc Res 2008;10(4):R65.

[181] de Macêdo Andrade AC, et al. Molecular breast cancer subtypes and therapies in a public hospital of Northeastern Brazil. BMC Women's Health 2014;14(1):110.

[182] Molina R, et al. Tumor markers in breast cancer−European group on tumor markers recommendations. Tumor Biol 2005;26(6):281−93.

[183] Wang Z, et al. CD44 directed nanomicellar payload delivery platform for selective anticancer effect and tumor specific imaging of triple negative breast cancer. Nanomed Nanotechnol Biol Med 2018;14(4):1441−54.

[184] Watson SS, et al. Microenvironment-mediated mechanisms of resistance to HER2 inhibitors differ between HER2+ breast cancer subtypes. Cell Systems 2018;6(3):329−42. e6.

[185] Fowler AM, Linden HM. Functional estrogen receptor imaging before neoadjuvant therapy for primary breast cancer. J Nucl Med 2017;58(4):560−2.

[186] Zhang C, et al. High F-content perfluoropolyether-based nanoparticles for targeted detection of breast cancer by 19F magnetic resonance and optical imaging. ACS Nano 2018;12(9):9162−76.

[187] Kievit FM, et al. Targeting of primary breast cancers and metastases in a transgenic mouse model using rationally designed multifunctional SPIONs. ACS Nano 2012;6(3):2591−601.

[188] Zhao Y, et al. Gold nanoclusters doped with 64Cu for CXCR4 positron emission tomography imaging of breast cancer and metastasis. ACS Nano 2016;10(6):5959−70.

[189] Balasubramanian P, et al. Antibody-independent capture of circulating tumor cells of non-epithelial origin with the ApoStream® system. PloS One 2017;12(4):e0175414.

[190] Eghtedari M, et al. Engineering of hetero-functional gold nanorods for the in vivo molecular targeting of breast cancer cells. Nano Lett 2008;9(1):287−91.

[191] Herbst RS, Morgensztern D, Boshoff C. The biology and management of non-small cell lung cancer. Nature 2018;553(7689):446.

[192] Travis WD, Brambilla E, Riely GJ. New pathologic classification of lung cancer: relevance for clinical practice and clinical trials. J Clin Oncol 2013;31(8):992−1001.

[193] Lemjabbar-Alaoui H, et al. Lung cancer: biology and treatment options. Biochim Biophys Acta Rev Canc 2015;1856(2):189−210.

[194] Robinson CG, Bradley JD. The treatment of early-stage disease. In: Seminars in radiation oncology. Elsevier; 2010.

[195] Shtivelman E, et al. Molecular pathways and therapeutic targets in lung cancer. Oncotarget 2014;5(6):1392.

[196] Lynch TJ, et al. Activating mutations in the epidermal growth factor receptor underlying responsiveness of non−small-cell lung cancer to gefitinib. N Engl J Med 2004;350(21):2129−39.

[197] Pao W, et al. EGF receptor gene mutations are common in lung cancers from "never smokers" and are associated with sensitivity of tumors to gefitinib and erlotinib. Proc Natl Acad Sci USA 2004;101(36):13306—11.

[198] Paez JG, et al. EGFR mutations in lung cancer: correlation with clinical response to gefitinib therapy. Science 2004;304(5676):1497—500.

[199] Cohen AS, et al. Cell-surface marker discovery for lung cancer. Oncotarget 2017;8(69):113373.

[200] Zhao Y, et al. Anti-EGFR peptide-conjugated triangular gold nanoplates for computed tomography/photoacoustic imaging-guided photothermal therapy of non-small cell lung cancer. ACS Appl Mater Interfaces 2018;10(20):16992—7003.

[201] Ehlerding EB, et al. ImmunoPET imaging of CTLA-4 expression in mouse models of non-small cell lung cancer. Mol Pharm 2017;14(5):1782—9.

[202] England CG, et al. ImmunoPET imaging of CD146 in murine models of intra-pulmonary metastasis of non-small cell lung cancer. Mol Pharm 2017;14(10): 3239—47.

[203] Cohen AS, et al. Delta-opioid receptor (δOR) targeted near-infrared fluorescent agent for imaging of lung cancer: synthesis and evaluation in vitro and in vivo. Bioconjugate Chem 2015;27(2):427—38.

[204] Kuipers EJ, et al. Colorectal cancer. Nature Rev Dis Prim 2015;1:15065.

[205] Chand M, et al. Novel biomarkers for patient stratification in colorectal cancer: a review of definitions, emerging concepts, and data. World J Gastrointest Oncol 2018;10(7):145.

[206] Ahlquist DA, et al. Stool DNA and occult blood testing for screen detection of colorectal neoplasia. Ann Intern Med 2008;149(7):441—50.

[207] Yiu AJ, Yiu CY. Biomarkers in colorectal cancer. Anticancer Res 2016;36(3): 1093—102.

[208] Makris NE, et al. PET/CT-derived whole-body and bone marrow dosimetry of 89Zr-cetuximab. J Nucl Med 2015;56(2):249—54.

[209] Toy R, et al. Targeted nanotechnology for cancer imaging. Adv Drug Deliv Rev 2014;76:79—97.

[210] Nakase H, et al. Evaluation of a novel fluorescent nanobeacon for targeted imaging of Thomsen-Friedenreich associated colorectal cancer. Int J Nanomed 2017;12:1747.

[211] Jeong S, et al. Cancer-microenvironment-sensitive activatable quantum dot probe in the second near-infrared window. Nano Lett 2017;17(3):1378—86.

[212] Kwon OS, et al. Dual-color emissive upconversion nanocapsules for differential cancer bioimaging in vivo. ACS Nano 2016;10(1):1512—21.

[213] Beack S, et al. Hyaluronate—peanut agglutinin conjugates for target-specific bio-imaging of colon cancer. Bioconjugate Chem 2017;28(5):1434—42.

[214] Burggraaf J, et al. Detection of colorectal polyps in humans using an intravenously administered fluorescent peptide targeted against c-met. Nat Med 2015;21(8):955.

[215] Si HY, et al. Carboxylate-containing two-photon probe for simultaneous detection of extra-and intracellular pH values in colon cancer tissue. Anal Chem 2018;90(13): 8058—64.

[216] Sathianathen NJ, et al. Landmarks in prostate cancer. Nat Rev Urol 2018:1.

[217] Hajipour H, et al. Enhanced anti-cancer capability of ellagic acid using solid lipid nanoparticles (SLNs). Int J Canc Manag 2018;11(1).

[218] Chistiakov DA, et al. New biomarkers for diagnosis and prognosis of localized prostate cancer. In: Seminars in cancer biology. Elsevier; 2018.

[219] Li R, et al. The use of PET/CT in prostate cancer. Prostate Cancer & Prostatic Dis 2017:1.

[220] Agemy L, et al. Nanoparticle-induced vascular blockade in human prostate cancer. Blood 2010;116(15):2847—56. blood-2010-03-274258.

[221] Kelly KA, et al. Detection of early prostate cancer using a hepsin-targeted imaging agent. Canc Res 2008;68(7):2286—91.

[222] Cai H, et al. Bombesin functionalized 64Cu-copper sulfide nanoparticles for targeted imaging of orthotopic prostate cancer. Nanomedicine 2018;(0).

[223] Yeh C-Y, et al. Peptide-conjugated nanoparticles for targeted imaging and therapy of prostate cancer. Biomaterials 2016;99:1—15.

[224] Hu H, et al. Dysprosium-modified tobacco mosaic virus nanoparticles for ultra-high-field magnetic resonance and near-infrared fluorescence imaging of prostate cancer. ACS Nano 2017;11(9):9249—58.

[225] Siegel RL, Miller KD, Jemal A. Cancer statistics. CA Cancer J Clin 2015;65(1):5—29.

[226] Singh D, et al. Recent advances in pancreatic cancer: biology, treatment, and prevention. Biochim Biophys Acta Rev Canc 2015;1856(1):13—27.

[227] Lee CJ, Heidt DG, Simeone DM. Pancreatic cancer stem cells. In: Encyclopedia of cancer. Springer; 2008. p. 2254—7.

[228] Patra CR, et al. Fabrication of gold nanoparticles for targeted therapy in pancreatic cancer. Adv Drug Deliv Rev 2010;62(3):346—61.

[229] Kleeff J, et al. Pancreatic cancer. Nat Rev Dis Primers 2016;2:16022.

[230] Makohon-Moore A, Iacobuzio-Donahue CA. Pancreatic cancer biology and genetics from an evolutionary perspective. Nat Rev Canc 2016;16(9):553.

[231] Zhou H, et al. IGF1 receptor targeted theranostic nanoparticles for targeted and image-guided therapy of pancreatic cancer. ACS Nano 2015;9(8):7976—91.

[232] Jiang Y, et al. Magnetic mesoporous nanospheres anchored with LyP-1 as an efficient pancreatic cancer probe. Biomaterials 2017;115:9—18.

[233] Zettlitz KA, et al. Dual-modality immuno-PET and near-infrared fluorescence imaging of pancreatic cancer using an anti—prostate stem cell antigen cys-diabody. J Nucl Med 2018;59(9):1398—405.

[234] Kim H-Y, et al. RAGE-specific single chain Fv for PET imaging of pancreatic cancer. PloS One 2018;13(3):e0192821.

[235] Ferlay J, et al. Estimates of worldwide burden of cancer in 2008: GLOBOCAN 2008. Int J Canc 2010;127(12):2893—917.

[236] Czerniak B, Dinney C, McConkey D. Origins of bladder cancer. Annu Rev Pathol 2016;11:149—74.

[237] Knowles MA, Hurst CD. Molecular biology of bladder cancer: new insights into pathogenesis and clinical diversity. Nat Rev Canc 2015;15(1):25.

[238] Damrauer JS, et al. Intrinsic subtypes of high-grade bladder cancer reflect the hallmarks of breast cancer biology. Proc Natl Acad Sci USA 2014;111(8):3110—5.

[239] Sjödahl G, et al. A molecular taxonomy for urothelial carcinoma. Clinical cancer research; 2012.

[240] Choi W, et al. Identification of distinct basal and luminal subtypes of muscle-invasive bladder cancer with different sensitivities to frontline chemotherapy. Canc Cell 2014;25(2):152—65.

[241] Aine M, et al. On molecular classification of bladder cancer: out of one, many. Eur Urol 2015;68(6):921—3.

[242] Pan Y, et al. Endoscopic molecular imaging of human bladder cancer using a CD47 antibody. Sci Transl Med 2014;6(260). 260ra148-260ra148.

[243] Nishizawa K, et al. Fluorescent imaging of high-grade bladder cancer using a specific antagonist for chemokine receptor CXCR4. Int J Canc 2010;127(5):1180—7.

[244] Davis RM, et al. Surface-enhanced Raman scattering nanoparticles for multiplexed imaging of bladder cancer tissue permeability and molecular phenotype. ACS Nano 2018;12(10):9669—79.

[245] Bray F, et al. Global cancer statistics 2018: GLOBOCAN estimates of incidence and mortality worldwide for 36 cancers in 185 countries. CA: a cancer journal for clinicians; 2018.

[246] Antuña AR, et al. Brain metastases of non—small cell lung cancer: prognostic factors in patients with surgical resection. J Neurol Surg Cent Eur Neurosurg 2018;79(02): 101—7.

[247] Mischel PS, Cloughesy TF, Nelson SF. DNA-microarray analysis of brain cancer: molecular classification for therapy. Nat Rev Neurosci 2004;5(10):782.

[248] Roose T, et al. Solid stress generated by spheroid growth estimated using a linear poroelasticity model☆. Microvasc Res 2003;66(3):204—12.

[249] Pasqualini R, et al. Aminopeptidase N is a receptor for tumor-homing peptides and a target for inhibiting angiogenesis. Canc Res 2000;60(3):722—7.

[250] Ni D, et al. Dual-targeting upconversion nanoprobes across the blood—brain barrier for magnetic resonance/fluorescence imaging of intracranial glioblastoma. ACS Nano 2014;8(2):1231—42.

[251] Huang N, et al. Efficacy of NGR peptide-modified PEGylated quantum dots for crossing the blood—brain barrier and targeted fluorescence imaging of glioma and tumor vasculature. Nanomed Nanotechnol Biol Med 2017;13(1):83—93.

[252] Li C, et al. Preoperative detection and intraoperative visualization of brain tumors for more precise surgery: a new dual-modality MRI and NIR nanoprobe. Small 2015;11(35):4517—25.

[253] Diaz RJ, et al. Focused ultrasound delivery of Raman nanoparticles across the blood-brain barrier: potential for targeting experimental brain tumors. Nanomed Nanotechnol Biol Med 2014;10(5):e1075—87.

[254] Marie H, et al. Superparamagnetic liposomes for MRI monitoring and external magnetic field-induced selective targeting of malignant brain tumors. Adv Funct Mater 2015;25(8):1258—69.

[255] Zheng M, et al. Self-targeting fluorescent carbon dots for diagnosis of brain cancer cells. ACS Nano 2015;9(11):11455—61.

[256] Xu HL, et al. Glioma-targeted delivery of a theranostic liposome integrated with quantum dots, superparamagnetic iron oxide, and cilengitide for dual-imaging guiding cancer surgery. Adv Healthc Mater 2018;7(9):1701130.

[257] Matulonis UA, et al. Ovarian cancer. Nat Rev Dis Primers 2016;2:16061.

[258] Reyners A, et al. Molecular imaging in ovarian cancer. Ann Oncol 2016;27(Suppl. l_1):i23—9.

[259] van Kruchten M, et al. Assessment of estrogen receptor expression in epithelial ovarian cancer patients using 16a-18F-fluoro-17bestradiol PET/CT. J Nucl Med 2015;56(1):50—5.

[260] Da Pieve C, et al. Efficient [18F] AlF radiolabeling of ZHER3: 8698 affibody molecule for imaging of HER3 positive tumors. Bioconjugate Chem 2016;27(8): 1839—49.

[261] Patel NR, et al. Design, synthesis, and characterization of folate-targeted platinum-loaded theranostic nanoemulsions for therapy and imaging of ovarian cancer. Mol Pharm 2016;13(6):1996—2009.

[262] Van Dam GM, et al. Intraoperative tumor-specific fluorescence imaging in ovarian cancer by folate receptor-α targeting: first in-human results. Nat Med 2011;17(10): 1315.

[263] Gao Y, et al. Ultrasound molecular imaging of ovarian cancer with CA-125 targeted nanobubble contrast agents. Nanomed Nanotechnol Biol Med 2017;13(7):2159—68.

[264] Williams RM, et al. Noninvasive ovarian cancer biomarker detection via an optical nanosensor implant. Sci Adv 2018;4(4):eaaq1090.

[265] Satpathy M, et al. Active targeting using her-2-affibody-conjugated nanoparticles enabled sensitive and specific imaging of orthotopic her-2 positive ovarian tumors. Small 2014;10(3):544—55.

[266] Kinsella JM, et al. X-ray computed tomography imaging of breast cancer by using targeted peptide-labeled bismuth sulfide nanoparticles. Angew Chem Int Ed 2011;50(51):12308—11.

[267] Zhang L, et al. Phosphatidylserine-targeted bimodal liposomal nanoparticles for in vivo imaging of breast cancer in mice. J Contr Release 2014;183:114—23.

[268] Xi L, et al. Molecular photoacoustic tomography of breast cancer using receptor targeted magnetic iron oxide nanoparticles as contrast agents. J Biophot 2014;7(6): 401—9.

[269] Das M, Duan W, Sahoo SK. Multifunctional nanoparticle—EpCAM aptamer bio-conjugates: a paradigm for targeted drug delivery and imaging in cancer therapy. Nanomed Nanotechnol Biol Med 2015;11(2):379—89.

[270] Sun X, et al. Molecular imaging of tumor-infiltrating macrophages in a preclinical mouse model of breast cancer. Theranostics 2015;5(6):597.

[271] Zhong Y, et al. Hyaluronic acid-shelled acid-activatable paclitaxel prodrug micelles effectively target and treat CD44-overexpressing human breast tumor xenografts in vivo. Biomaterials 2016;84:250—61.

[272] Ulaner GA, et al. Detection of HER2-positive metastases in patients with HER2-negative primary breast cancer using 89Zr-trastuzumab PET/CT. J Nucl Med 2016;57(10):1523.

[273] Jafari A, et al. Synthesis and characterization of Bombesin-superparamagnetic iron oxide nanoparticles as a targeted contrast agent for imaging of breast cancer using MRI. Nanotechnology 2015;26(7):075101.

[274] Derlin T, et al. Molecular imaging of chemokine receptor CXCR4 in non-small cell lung cancer using 68Ga-pentixafor PET/CT: comparison with 18F-FDG. Clin Nucl Med 2016;41(4):e204—5.

[275] Oh S-G, et al. In vivo visualization of the migration of mast cell toward lung cancer lesion using nuclear medicine imaging with sodium iodide symporter. J Nucl Med 2017;58(Suppl. 1). pp. 62—62.

[276] Ehlerding EB, et al. CD38 as a PET imaging target in lung cancer. Mol Pharm 2017;14(7):2400—6.

[277] Liu F, et al. Design, synthesis, and biological evaluation of 68Ga-DOTA—PA1 for lung cancer: a novel PET tracer for multiple somatostatin receptor imaging. Mol Pharm 2018;15(2):619—28.

[278] Kang L, et al. Noninvasive trafficking of Brentuximab vedotin and PET imaging of CD30 in lung cancer murine models. Mol Pharm 2018;15(4):1627—34.

[279] Liu S, et al. Design, synthesis, and validation of Axl-targeted monoclonal antibody probe for microPET imaging in human lung cancer xenograft. Mol Pharm 2014;11(11):3974—9.

[280] Berroterán-Infante N, et al. [18F] FEPPA: improved automated radiosynthesis, binding affinity, and preliminary in vitro evaluation in colorectal cancer. ACS Med Chem Lett 2018;9(3):177—81.

[281] Rezazadeh F, et al. 99m Tc-D (LPR): a novel retro-inverso peptide for VEGF receptor— 1 targeted tumor imaging. Nucl Med Biol 2018;62:54—62.

[282] Li X, et al. Evolution of DNA aptamers through in vitro metastatic-cell-based systematic evolution of ligands by exponential enrichment for metastatic cancer recognition and imaging. Anal Chem 2015;87(9):4941—8.

[283] Liang L, et al. Facile assembly of functional upconversion nanoparticles for targeted cancer imaging and photodynamic therapy. ACS Appl Mater Interfaces 2016;8(19):11945—53.

[284] Kim J, et al. Molecular imaging of colorectal tumors by targeting colon cancer secreted protein-2 (CCSP-2). Neoplasia 2017;19(10):805—16.

[285] Matsuzaki H, et al. Novel hexosaminidase-targeting fluorescence probe for visualizing human colorectal cancer. Bioconjugate Chem 2016;27(4):973—81.

[286] Tiernan JP, et al. CEA-targeted nanoparticles allow specific in vivo fluorescent imaging of colorectal cancer models. Nanomedicine 2015;10(8):1223—31.

[287] Rabinsky EF, et al. Overexpressed claudin-1 can be visualized endoscopically in colonic adenomas in vivo. Cellular & Mol Gastroenterol & Hepatol 2016;2(2): 222—37.

[288] Zhou B, et al. PET imaging of Dll4 expression in glioblastoma and colorectal cancer xenografts using 64Cu-labeled monoclonal antibody 61B. Mol Pharm 2015;12(10):3527—34.

[289] Rauscher I, et al. Intrapatient comparison of 111In-PSMA I&T SPECT/CT and hybrid 68Ga-HBED-CC PSMA PET in patients with early recurrent prostate cancer. Clin Nucl Med 2016;41(9):e397—402.

[290] Gao X, et al. In vivo cancer targeting and imaging with semiconductor quantum dots. Nat Biotechnol 2004;22(8):969.

[291] Lee C-M, et al. Prostate cancer-targeted imaging using magnetofluorescent polymeric nanoparticles functionalized with bombesin. Pharmaceut Res 2010;27(4):712—21.

[292] Lo S-T, et al. Dendrimer nanoscaffolds for potential theranostics of prostate cancer with a focus on radiochemistry. Mol Pharm 2013;10(3):793—812.

[293] Ghosh D, et al. M13-templated magnetic nanoparticles for targeted in vivo imaging of prostate cancer. Nat Nanotechnol 2012;7(10):677.

[294] Guo L, et al. Prostate cancer targeted multifunctionalized graphene oxide for magnetic resonance imaging and drug delivery. Carbon 2016;107:87—99.

[295] Sun Y, et al. Preclinical study on GRPR-targeted 68Ga-probes for PET imaging of prostate cancer. Bioconjugate Chem 2016;27(8):1857—64.

[296] Hong H, et al. Generation and screening of monoclonal antibodies for immunoPET imaging of IGF1R in prostate cancer. Mol Pharm 2014;11(10):3624—30.

[297] Persson M, et al. uPAR targeted radionuclide therapy with 177Lu-DOTA-AE105 inhibits dissemination of metastatic prostate cancer. Mol Pharm 2014;11(8):2796—806.

[298] Richter S, et al. Metabolically stabilized 68Ga-NOTA-Bombesin for PET imaging of prostate cancer and influence of protease inhibitor phosphoramidon. Mol Pharm 2016;13(4):1347—57.

[299] Kamerkar S, et al. Exosomes facilitate therapeutic targeting of oncogenic KRAS in pancreatic cancer. Nature 2017;546(7659):498.

[300] Montet X, Weissleder R, Josephson L. Imaging pancreatic cancer with a peptide— nanoparticle conjugate targeted to normal pancreas. Bioconjugate Chem 2006;17(4): 905—11.

[301] Huynh AS, et al. Novel toll-like receptor 2 ligands for targeted pancreatic cancer imaging and immunotherapy. J Med Chem 2012;55(22):9751—62.

[302] Qian J, et al. Imaging pancreatic cancer using surface-functionalized quantum dots. J Phys Chem B 2007;111(25):6969—72.

[303] Lee GY, et al. Theranostic nanoparticles with controlled release of gemcitabine for targeted therapy and MRI of pancreatic cancer. ACS Nano 2013;7(3):2078—89.

[304] Amirkhanov NV, et al. Imaging human pancreatic cancer xenografts by targeting mutant KRAS2 mRNA with [111In] DOTA n-poly (diamidopropanoyl) m-KRAS2 PNA-d (Cys-Ser-Lys-Cys) nanoparticles. Bioconjugate Chem 2010;21(4):731—40.

[305] Wang M, et al. The development of [18F] AlF-NOTA-NT as PET agents of neurotensin receptor-1 positive pancreatic cancer. Mol Pharm 2018;15(8):3093—100.

[306] Houghton JL, et al. Preloading with unlabeled CA19. 9 targeted human monoclonal antibody leads to improved PET imaging with 89Zr-5B1. Mol Pharm 2017;14(3):908–15.

[307] Zhang D, et al. High-performance identification of human bladder cancer using a signal self-amplifiable photoacoustic nanoprobe. ACS Appl Mater Interfaces 2018;10(34):28331–9.

[308] Yuan R, et al. Quantum dot-based fluorescent probes for targeted imaging of the EJ human bladder urothelial cancer cell line. Experiment & Therapeutic Med 2018;16(6):4779–83.

[309] Wang J, et al. Identification of carbonic anhydrase IX as a novel target for endoscopic molecular imaging of human bladder cancer. Cell Physiol Biochem 2018;47(4): 1565–77.

[310] Allen MD, et al. Prognostic and therapeutic impact of argininosuccinate synthetase 1 control in bladder cancer as monitored longitudinally by PET imaging. Canc Res 2014;74(3):896–907.

[311] Paquette M, et al. Targeting IL-5Rα with antibody-conjugates reveals a strategy for imaging and therapy for invasive bladder cancer. OncoImmunology 2017;6(10): e1331195.

[312] Dixit S, et al. Transferrin receptor-targeted theranostic gold nanoparticles for photosensitizer delivery in brain tumors. Nanoscale 2015;7(5):1782–90.

[313] Patil R, et al. MRI virtual biopsy and treatment of brain metastatic tumors with targeted nanobioconjugates: nanoclinic in the brain. ACS Nano 2015;9(5):5594–608.

[314] Luo H, et al. Noninvasive brain cancer imaging with a bispecific antibody fragment, generated via click chemistry. Proc Natl Acad Sci USA 2015;112(41):12806–11.

[315] Yan H, et al. Two-order targeted brain tumor imaging by using an optical/para-magnetic nanoprobe across the blood brain barrier. ACS Nano 2011;6(1):410–20.

[316] Tian T, et al. Targeted imaging of brain tumors with a framework nucleic acid probe. ACS Appl Mater Interfaces 2018;10(4):3414–20.

[317] Hao Y, et al. Targeted imaging and chemo-phototherapy of brain cancer by a multifunctional drug delivery system. Macromol Biosci 2015;15(11):1571–85.

[318] Hernandez R, et al. ImmunoPET imaging of CD146 expression in malignant brain tumors. Mol Pharm 2016;13(7):2563–70.

[319] Barua A, et al. Interleukin 16-(IL-16-) targeted ultrasound imaging agent improves detection of ovarian tumors in laying hens, a preclinical model of spontaneous ovarian cancer. BioMed Res Int 2015;2015.

[320] Barua A, et al. Enhancement of ovarian tumor detection by DR6-targeted ultrasound imaging agents in laying hen model of spontaneous ovarian cancer. Int J Gynecol Canc 2016;26(8):1375–85.

[321] Lutz AM, et al. Ultrasound molecular imaging in a human CD276 expression-modulated murine ovarian cancer model. Clin Canc Res 2014;20(5):1313–22.

[322] Munnink THO, et al. 89Zr-trastuzumab PET visualises HER2 downregulation by the HSP90 inhibitor NVP-AUY922 in a human tumour xenograft. Eur J Canc 2010;46(3):678–84.

[323] Bensch F, et al. Phase I imaging study of the HER3 antibody RG7116 using 89Zr-RG7116-PET in patients with metastatic or locally advanced HER3-positive solid tumors. American Society of Clinical Oncology; 2014.

[324] Prantner AM, et al. Molecular imaging of mesothelin-expressing ovarian cancer with a human and mouse cross-reactive nanobody. Mol Pharm 2018;15(4):1403–11.

[325] Guerrero Y, et al. Targeted imaging of ovarian cancer cells using viral nanoparticles doped with indocyanine green. In: Optical methods for tumor treatment and detection: mechanisms and techniques in photodynamic therapy XXII. International Society for Optics and Photonics; 2013.

[326] Perrone MG, et al. PET radiotracer [18F]-P6 selectively targeting COX-1 as a novel biomarker in ovarian cancer: preliminary investigation. Eur J Med Chem 2014;80:562—8.

[327] Makvandi M, et al. A PET imaging agent for evaluating PARP-1 expression in ovarian cancer. J Clin Invest 2018;128(5):2116—26.

[328] Park S-m, et al. Towards clinically translatable in vivo nanodiagnostics. Nat Rev Mater 2017;2(5):17014.

[329] ter Weele EJ, et al. Development, preclinical safety, formulation, and stability of clinical grade bevacizumab-800CW, a new near infrared fluorescent imaging agent for first in human use. Eur J Pharm Biopharm 2016;104:226—34.

[330] Benezra M, et al. Multimodal silica nanoparticles are effective cancer-targeted probes in a model of human melanoma. J Clin Invest 2011;121(7):2768—80.

[331] Willmann JK, et al. Ultrasound molecular imaging with BR55 in patients with breast and ovarian lesions: first-in-human results. J Clin Oncol: Off J Am Soc Clin Oncol 2017;35(19):2133—40.

CHAPTER 5

Challenges and future directions

Currently, cancer patients are categorized based on the site and tissue of origin of the disease. However, it is becoming increasingly clear that the biological barriers and heterogeneity seen in tumors and patients call for more targeted therapy than found in conventional cancer treatment methods. Cancer is a worldwide health issue, and there is an urgent need to identify more effective and noninvasive biomarkers for early diagnosis, prognosis, and therapeutic targeting based on individual patient characteristics. First, some challenges in targeted delivery are discussed, and then several relatively new approaches that may be used to facilitate imaging for cancer detection and treatment are described.

1. Challenges of targeted delivery

The development of targeted cargo delivery and active pharmaceutical agent for specific disease pathologies, that is, the process of drug discovery, has been instrumental in generating new lead molecules for a plethora of disease conditions. The serendipitous success of many early drugs led to nearsighted evaluations of their success. Nevertheless, their discovery was expedited by the introduction of novel tools, including high-throughput screening, structure–activity relationships, combinatorial chemistry, computer-aided drug design, and artificial intelligence. Although the process of lead identification is robust, drugs often fail in later stages of development, typically due to safety and efficacy concerns that fundamentally arise from high accumulation in off-target organs or poor accumulation in target organs, respectively [1]. This has been a major roadblock for the translation of potent drug candidates, which inherently possess excellent potential but fail to demonstrate significant clinical impact because of dose-related toxicities or dose-limited efficacies due to off-target effects. Drug delivery technologies hold the potential to address this limitation and have emerged in parallel to the drug discovery process. Over the years, drug delivery research has offered multiple approaches to target drugs, including local therapies such as topical formulations and

Targeted Cancer Imaging
ISBN 978-0-12-824513-2
https://doi.org/10.1016/B978-0-12-824513-2.00006-1

physical devices. Local therapies offer the simplest means of targeting but are not practical when disease sites are hard to reach. Nanoparticles (NPs) have been developed to target therapies to specific tissues, and the modulation of physicochemical properties such as size and charge could improve NP targeting of specific tissues. However, NPs face biological barriers that impede their targeting capabilities. The advent of nucleic acid-based therapies, gene therapies, and cell therapies has opened additional therapeutic opportunities but has also introduced new delivery challenges. Several reviews exist on ligand-targeted drug delivery systems, with some of them dealing particularly with the chemical and biochemical aspects of targeted constructs.

2. The nanoparticle journey

NP delivery systems interact with various organs, tissues, cells, and molecules as they are transported throughout the body (Fig. 5.1). The relationships

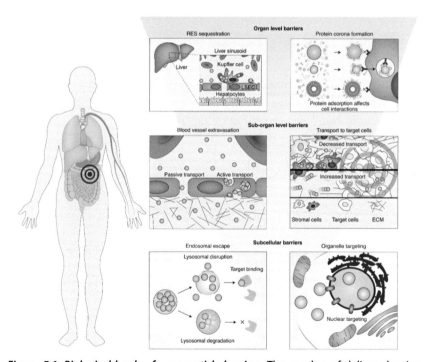

Figure 5.1 *Biological levels of nanoparticle barriers.* The number of delivery barriers increases with a deeper delivery target. Organs are typically the easiest to deliver nanomaterials to, while subcellular structures are the most difficult because the NPs have more barriers to travel through to get to the final destination [26].

between engineered nanomaterials and biological systems are referred to as nanoparticle—biological interactions. These interactions dictate what happens to the nanomaterial inside the body. When NPs are administered, proteins instantly adsorb to the NP surface and form a protein corona. Protein corona (PC) profile, composition, and assembly depend on the biological molecules and fluids the NP first interacts with and physical and chemical properties of NPs, such as size, shape, and surface charge [2]. This corona forms a new interface between the NPs and cells or tissues and influences NP uptake, biodistribution, and immune response [3,4]. The PC may alter the in vivo trajectory of NPs and can mask the targeting effect of engineered ligands on NP surfaces. Salvati et al. understood that cell-specific targeting with transferrin-conjugated NPs decreased when NPs were coated in serum proteins compared with when they were not [5]. The majority of NPs circulating in the blood are typically removed by the liver and spleen of the reticuloendothelial system (RES). The function of these organs is to filter blood and remove biological debris and foreign particulates from circulation. These organs form a significant barrier to intravenously administered NPs, including micelles, quantum dots (QDs), liposomes, and gold NPs [6,7]. The liver can sequester the majority of intravenously administered NPs and retain nondegradable NPs for months after administration [8]. The primary source of NP sequestration is Kupffer cells. These phagocytic immune cells line the inside of liver sinusoids and capture NPs as they pass by in circulation. Tsoi et al. demonstrated that sinusoids slow blood velocity compared with arteries and veins, increasing the probability of NP interactions and uptake by Kupffer cells [9]. This can make the RES system a suitable target for intravenously administered nanomedicines [10], as they will largely accumulate there. However, avoidance of the RES is essential for improving the delivery efficiency of intravenously administered nanoparticles to targets outside of the RES [11]. Other organs will also remove NPs depending on their physical and chemical properties, such as their size. For example, NPs smaller than 6 nm can be renally excreted by the kidneys, and intradermally administered NPs can be sequestered by dendritic cells [12]. These are just two examples of the different cells and tissues that remove NPs from circulation and prevent them from reaching the biological target site. NPs that reach the target organ must exit the vasculature to reach target cells within the tissue. NPs transport through the blood vessel is dependent on the vessel physiology. For example, the vessels of the liver sinusoid are fenestrated, so NPs smaller than the fenestrae (approximately <100 nm) can diffuse through to access the space of Disse. The vessels in the glomeruli of the kidney also have fenestrae. The effective cut-off size is smaller (<6 nm) due to the structure and composition of the underlying glomerular basement membrane [13,14]. NP size, surface charge, and shape can affect NP clearance by the kidneys. The vessels in the brain are tightly regulated by the blood-brain barrier, which prevents the delivery of NPs carrying drugs or imaging agents to the brain [15]. The vessels of solid tumors use a combination of active and passive transport mechanisms to transport NPs. Vessel physiology varies across different endothelial linings. The physicochemical properties of NPs should be designed to extravasate through blood vessels at target tissues. NPs inside a target tissue must travel through the tissue stroma to reach target cells. The tissue stroma includes extracellular matrix (ECM) and connective tissue cells, such as fibroblasts, pericytes, and tissue-specific support cells. ECM proteins can trap NPs before they reach their intended target. The composition of the ECM varies between tissues and can become drastically altered in diseases such as liver fibrosis and cancer [16]. Some ECM components, such as collagen, fibrinogen, and hyaluronic acid, can sterically hinder nanoparticle diffusion [17]. Off-target cells present within the stroma can sequester NPs before they reach their intended target cell type. For example, tumor-associated macrophages can sequester NPs that reach the tumor, preventing them from delivering their cargo to cancer cells [18]. NPs must navigate through the tissue stroma and avoid sequestration or degradation in the ECM or by off-target cells in order to reach their target. NPs may need to enter the target cell in order for their cargo to elicit a therapeutic effect. Cellular uptake can occur through various mechanisms, such as membrane fusion, caveolin-mediated endocytosis, clathrin-mediated endocytosis, macropinocytosis, or phagocytosis [19]. Target cell phenotype and nanomaterial chemical composition determine the uptake and processing of nanoparticles. Surface receptor identities, expression levels, and recycling kinetics affect the uptake rate and pathways accessible to the NPs. This affects the optimal uptake route for a given cell type. Santos et al. found that 132N1 cells mainly used clathrin-mediated endocytosis while A549 cells mainly used the caveolin pathway [20]. The physicochemical characteristics of the nanomaterial, such as size, shape, and surface chemistry, also influence the uptake route. Meng et al. demonstrated that the uptake of silica NPs by HeLa and A549 cancer cells via macropinocytosis was ~40-fold greater when the aspect ratio was 2.1−2.5 versus other aspect ratios [21]. These examples that a variety of cell-type-specific mechanisms dictate NP−cell interactions and uptake. NPs inside the cell may need to escape the endosome to reach a final subcellular target location, such as the cytoplasm.

Strategies include the use of positively charged lipids to disrupt endosome bilayer stability or pH-sensitive polymers to modulate proton transport in endosomes [22]. In one application of this strategy, Hu et al. developed a pH-responsive polymer NP for cytosolic drug delivery. This NP responded to acidic lysosomes by increasing in diameter from 200 to 500 nm when the pH dropped from 7.4 to 4.9 to disrupt the membranes of acidic lysosomes. This translated to a ~16-fold increase in cytosolic localization of the delivered cargo in dendritic cells in vitro compared with a non-pH-responsive design [23]. Some drugs may need to access organelles within the cell. Pan et al. demonstrated improved nuclear delivery of silica nanoparticles to HeLa cancer cells by conjugating the HIV-TAT nuclear localization peptide to the nanoparticle surface [24]. This nuclear localization sequence binds the α and β importin receptors for active transport into the nucleus [25]. Methods allowing endosomal escape and using specific organelle localization tags should be considered for NPs targeting subcellular locations. Each level of nano/bio interactions has barriers that prevent NPs from being delivered to the target site. NPs can be sequestered or degraded at every barrier, reducing the number of NPs on the delivery journey to the target. Understanding interactions between NPs and the biology at each level will help to design efficient NCs optimized for the biology of the delivery pathway.

3. Removing barriers from the nanoparticle journey

Delivery efficiency can be defined as the percentage of administered NPs delivered to the intended biological target. It depends on the number of biological barriers and how the nanoparticles interact with them (Fig. 5.2A and C). Using tumor targeting as an example, the number of nanoparticles decreases as they move through the barriers (Fig. 5.2B). Conceptually, this is exemplified by reducing accumulation as they transport from the whole tumor to the cancer cell nucleus. Dai et al. showed that 0.7% of gold nanoparticles are delivered to the solid tumor, but only 0.0014% are delivered to the tumor cells in mouse models [27]. This is due to the biological barriers that an NP must overcome to go from the tumor vessel to the final target cell. The number of barriers can affect delivery efficiency. One approach to reduce the number of barriers is to change the administration route to directly bypass certain barriers. Many nanomedicines use intravenous delivery, which is suitable for targeting hematological, vascular, or systemically disseminated diseases since there is direct access to these sites

a

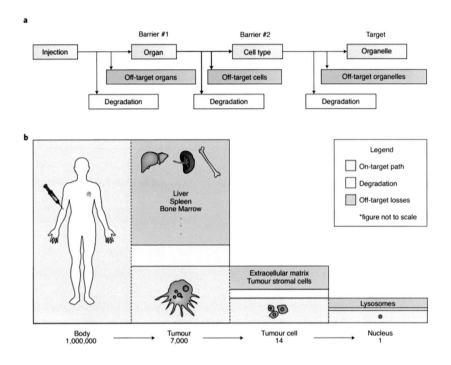

b

c

Figure 5.2 *A systematic view of nanoparticle delivery barriers.* (A) schematic proposing a barrier framework of the NP delivery process. Successive barriers in the body remove the administered dose of NP until only a small percentage is delivered to the intended target; (B) an example of using the barrier framework to model NP drug

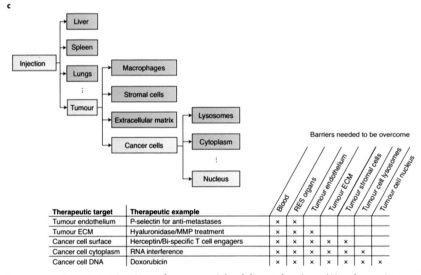

via blood circulation. Oral administration allows access to the gastrointestinal tract, intraocular administration to the eye, inhalation to the lungs, and intradermal to the skin and lymphatic system. Garbuzenko et al. showed that inhalation improved both the delivery to the lungs and the therapeutic efficacy compared with intravenous administration [28]. Intravenously administered NPs must pass the barriers of the RES system and face degradation in circulation before reaching the lung. Inhaled NPs have more direct access to the lungs and face fewer delivery barriers. Another approach is to change the biological target. If the disease of interest has multiple druggable targets, the most direct delivery pathway may favor higher delivery efficiency. In the example of treating a solid tumor in Fig. 5.2C, targeting the tumor endothelium has fewer barriers to overcome compared with targeting the nuclei of tumor cells. Clinically approved nanomedicines use this concept. They target pathologies located in tissues where nanoparticles preferentially accumulate, such as the skin and RES. Examples include Doxil (liposomal doxorubicin) for the treatment of AIDS-related Kaposi sarcoma skin tumors and Feraheme (carbohydrate-coated iron oxide nanoparticle) for treating iron-deficiency anemia in the liver and spleen [29].

4. Identifying the ideal nanoparticle design

Collecting nano/bio interaction data. A significant number of studies aim to elucidate the relationship between the physicochemical properties of engineered nanomaterials and their interaction with biological systems in vitro and in vivo. However, it is a complex set of interactions that are unlikely to be defined by a single parameter. (1) Data from in vitro studies can be used to understand cellular or subcellular level nano—bio interactions. This is exemplified by in vitro studies investigating nanoparticle

◀──────────────────────────────────────

delivery to the nucleus of a tumor. After the NPs are administered intravenously, they have to transport through the bloodstream to get to the final target site. Many of these NPs are taken up by the liver, spleen, and other reticuloendothelial organs. Once they enter the tumor, they have to cross the blood vessel, extracellular matrix, and other nontumor cells before reaching the tumor cells. Then they would have to cross the cell membrane, vesicles, and other subcellular structures before reaching the target in the nucleus. To illustrate how challenging this is, show that one out of one million nanoparticles may reach the nucleus with the successive loss of NPs along the delivery pathway. (C) a detailed description of the different barriers NPs must overcome to reach various therapeutic target locations for cancer therapy [26].

uptake by cells where essentially every property of the nanoparticle has an impact on cell uptake, including size, shape, ligand density, material composition, and surface chemistry [30,31]. Bai et al. showed that palladium and gold NPs were taken up more than platinum NPs. Wang et al. showed that the ligand valency of NP surfaces impacts their cellular interactions with SK-BR-3 and MCF-7 breast cancer cell lines [32]. (2) Data from in vivo studies are critical to delineate the role of multiple organs or systems in the delivery process. NP libraries can be created and administered to animals to understand how different NP properties affect certain biological outcomes. NP elimination from the body is one example where NP libraries have been used. Choi et al. established the 6 nm cut-off size for renal elimination using a library of different-sized quantum dots [13]. Poon et al. used a library of gold nanoparticles larger than 6 nm to show that NPs in this size range are eliminated through the hepatobiliary pathway in the feces or retained in the liver long-term [33]. Collection of nano/bio interaction data at both in vitro and in vivo levels is required to determine the optimal design. NP interactions at the disease site are also important to understand. The physiology of the tissue at the disease site will be different than in the healthy tissue and can affect NP delivery to the target cells. This has been studied in solid tumors where nanoparticle penetration and distribution are affected by collagen density, blood vessel density, blood vessel perfusion, and immune cell composition. Sykes et al. measured the collagen content of solid tumors and then modeled nanoparticle diffusion through collagen gels at different collagen densities [34,35]. They found that larger NPs (>60 nm) had reduced diffusion through higher collagen densities (>4 mg/mL) compared with smaller sizes and lower collagen densities. Ekdawi et al. measured vascular properties of solid tumors from mice injected intravenously with fluorescent liposomes [36]. They found that the liposomes accumulated in areas of the tumor with high vascular density, such as the periphery of the tumor. Stirland et al. investigated the role of vascular perfusion on local nanoparticle accumulation within tumors by injecting two uniquely fluorescent nanoparticles either sequentially or together and analyzing histology sections of the tumor [35]. They found that co-injected nanoparticles colocalized in the same parts of the tumor. When the two formulations were injected at different times, they accumulated in different areas of the tumor because the local blood vessel perfusion was dynamic. Cuccarese et al. investigated the impact of immune cell populations on nanoparticle accumulation by injecting lung tumor-bearing mice with fluorescent nanoparticles and then imaging the whole lungs for macrophage and nanoparticle distribution [37]. They found that the number of nanoparticles in a tumor correlated with the number of macrophages in that tumor. Furthermore, identifying the ideal nanoparticle formulation is complicated by pathophysiological changes in disease states. The presence of disease can alter the in vivo nano/bio interactions and change the nanoparticle's blood clearance properties. Kai et al. determined that nanoparticle clearance from the blood is faster in tumor-bearing mice than in healthy mice [38]. Similarly, Wu et al. showed in human patients that liposomal drugs were eliminated 1.5-fold more quickly in patients with liver tumors than patients without liver tumors [39]. Collecting data on nano—bio interactions is the key step in determining the optimal design. The abundance of data will likely require computational analysis to identify how the complex relationships between the nanomaterial and biology allow for identifying the optimal design.

5. Computational techniques to process nano/bio interaction data

Identifying the best nanoparticle design can be aided by experimentally examining how molecules, cells, and tissues interact with nanoparticles of different designs. Many of the examples in the preceding sections were focused on understanding how a single parameter contributes to a single biological outcome. However, within the body, there are many confounding and complex interactions between the nanomaterial and molecules, cells, and tissues that are not well understood. As the number of nanoparticle designs being tested increases and the amount of biological data collected about the nano/bio interactions increases, establishing complex relationships between these variables becomes possible. Computational techniques may be used to define the relationship between the nanomaterial properties and their biological interactions. Fig. 5.3 shows a general, high-level overview of this framework. This is an emerging area of research, and this section aims to highlight examples of computational approaches for understanding nano/bio interactions.

Linear regression models. Linear regression models can be used to estimate the relationship between a dependent variable and one or more independent variables. The advantage of this method is that it is simple to implement and evaluate. The main disadvantage is that it does not accurately model nonlinear relationships. Walkey et al. developed a partial least-squares regression model to predict cell association based on the PC

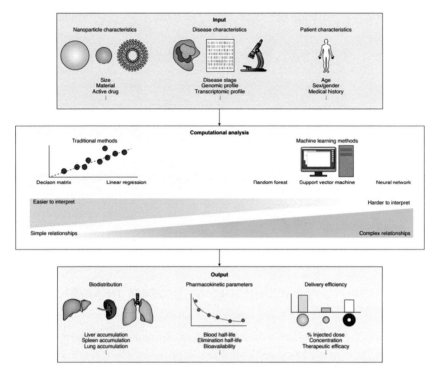

Figure 5.3 *A computational framework for analyzing nanoparticle—biological inter-action datasets.* Computational tools for understanding or predicting nanoparticle—biological interactions are useful for analyzing large datasets containing information about libraries of different nanoparticles and their biological interactions. There are nanoparticle properties (size, shape, surface chemistry, material, surface charge) and disease or patient-specific properties (disease stage, location in the body, phenotype, genomic profile, and age) that impact the ability of a given formulation to reach the target disease site. Statistics-based, computational approaches provide tools that can be used to correlate multiple input parameters to the desired output parameters such as the biodistribution, pharmacokinetic profile, or delivery efficiency of a given nanomaterial formulation to a specific delivery target [26].

of a panel of nanoparticle formulations [2]. A library of 105 different nanoparticle formulations (different sizes, materials, and surface chemistries) was incubated with serum, and the PC was measured using liquid chromatography-tandem mass spectrometry (LC-MS/MS). This data was used to generate a multivariable model that could predict cell association with 84% higher accuracy than single variable models using a single protein. This technique is typically used to establish the interaction of nanoparticles with simple biological systems such as cells in vitro.

Decision matrices. A decision matrix is a flow chart or table that can be used to identify key nanomaterial design parameters. Sykes et al. created a decision matrix to identify the optimal nanoparticle size for tumor accumulation for imaging (best contrast) or treatment (high retention and even distribution) applications based on tumor accumulation, fluorescence intensity, uptake rates, penetration capacity, and theoretical loading capacity data that were ranked in importance for therapeutic and diagnostic applications. Each nanoparticle size was scored for its usefulness in imaging or treating tumors of different sizes [40]. Poon et al. created a decision flowchart to identify nanoparticle elimination pathways based on size and chemical composition [33]. This can be used to decide which nanoparticle design to use to access different elimination pathways. The advantage of decision matrices is that they are simple to implement and understand. The disadvantage is that they guide decisions for only a few simple design parameters.

Machine learning. Researchers have begun exploring machine learning methods to develop predictive models with limited knowledge of the relationships between the large numbers of variables (such as support vector machines, neural networks, and random forests). Machine learning models are statistics–based computer algorithms that learn to perform a task without explicit instructions. When the relationship between a set of input variables (for example, nanoparticle characteristics) and a set of output variables (for example, tumor delivery) are unknown, machine learning can optimize a set of mathematical rules (that is, a model) to predict this relationship. Machine learning has been explored for applications such as to predict cell binding or accumulation in animals from the PC on nanoparticles, predict delivery to metastatic tumors [34,41,42], predict nanoparticle toxicity [43,44], and analyses chemical reagents for tuning nanoparticle physicochemical properties [45,46]. This usually requires hundreds to thousands of examples of accurately annotated data to train and validate the algorithm.

Support vector machines. Support vector machines are a method of supervised learning that can be used for classification or regression. Support vector machines are useful in situations with limited datasets (a few hundred data points) and when the prediction needs to tolerate noise in the dataset. Support vector machine models have been created to predict nanoparticle delivery to the metastatic tumor based on tumor 3D morphology [34] and to predict cell binding to different nanoparticle formulations from their PCs [47]. Kingston et al. built a model to predict nanoparticle delivery to metastatic tumors by training a support vector machine model from imaging data. Light-sheet microscopy was used to measure tumor morphology (that is, volume, surface area, sphericity, number of cells, cellular density, cell distance to blood vessels) and nanoparticle delivery (that is, number of nanoparticle positive cells, mean nanoparticle intensity) of over 1300 individual metastatic tumors. The support vector machine model was trained to predict the number of positive nanoparticle cells (output) from the morphological data about the tumor (inputs). The model was able to predict the number of positive nanoparticle cells with a Pearson correlation (r) of 0.94 and root mean squared error of 27 cells. This proof-of-concept study demonstrates how disease physiology can be used to predict nanoparticle delivery.

Neural networks. Neural networks work by relating input data to output results using linear and nonlinear transformations. The relationships are strengthened by large datasets iterating through multiple hidden layers of transformations. Each iteration adjusts the transformations, which improves prediction accuracy. The ability to adjust the number, connectivity and data transformations throughout the layers of the network makes these methods capable of modeling complex relationships between the dependent and independent variable(s). Lazarovits et al. used neural networks to generate a computational model to predict organ accumulation and blood half-life using PC compositions from a panel of nanoparticle formulations [41]. Different gold nanoparticle sizes were injected into rats. Their PCs at different time points were used as input data. The output data was the amount of gold in the liver, spleen, and blood. The neural network was trained to predict the number of gold nanoparticles in the liver, spleen, and blood based on the PC composition. When validated using two gold nanoparticle formulations of unknown composition, the model was able to predict the half-life, spleen accumulation, and liver accumulation based on the PC composition with 77%—94% accuracy. This study demonstrated a proof-of-concept that the PC composition could enable the prediction of nanoparticle delivery in animals. Random forests are an ensemble machine learning method for creating classification or regression models. Ban et al. used a random forest algorithm to predict the PC of nanoparticles given their physicochemical properties (size, shape, material, surface chemistry, surface charge) [42]. Data from the literature were collected from 652 different nanoparticles covering 40 different nanoparticle materials and over

50 types of surface modifications. Using 10-fold cross-validation, the random forest algorithm predicted the relative protein abundance of 178 different proteins, with most R-squared values >0.7 and root mean squared errors below 5%. In the future, the prediction of the PC from the physicochemical properties of the nanomaterial could offer a way to screen nanoparticle interactions computationally.

5.1 Tumor heterogeneity and ecological and evolutionary features of solid tumors

Genomic instability promotes cancer cells to undergo evolutionary dynamic under the constant selective pressure of harsh dynamic microenvironment that led to forming a variety of heterogeneous tumor cells (Fig. 5.4). During this process, hallmarks of tumors are formed. Dynamic evolution also drives the clonal selection procedure to produce committed metastasis cells for seeding in distant organs. A deep understanding of tumor heterogeneity accelerates the recognition of molecular events involved in metastasis, drug resistance, tumor recurrence and even encourages reformulation of conventional chemotherapeutics drugs. Indeed, clinical evaluation of tumor heterogeneity is an emergent issue to improve clinical oncology. In particular, intratumor heterogeneity is closely related to cancer progression, resistance to therapy, and recurrence. The different pattern of heterogeneity in tumor lead to the different treatment outcome and imply this concept that "no one size fit for all." At this time, personalized medicine is proposed as a solution for this problem. The tumor cell heterogeneity that is currently present in a population defines its capacity for therapy response. The relationship between genetic diversity and clinical outcomes is not universally consistent across different cancer types and can be complicated. Compared with homogeneous neoplasms, diverse neoplasms are more likely to harbor resistant clones and are more likely to evolve resistance in the future [48].

Tumor heterogeneity can be inter- and intrapatient, inter- and intratumor, and even appears at cellular levels such as cellular behavior and morphology and at subcellular levels such as genome regulation, genome expression, protein translation, and modification and cell signaling pathways (Fig. 5.5). For better understanding in the following sections, tumor heterogeneity is divided into two parts as "Evolutionary" and "Ecological" features of solid tumors [49,50]. Previously these indexes were recommended to be included in pathology reports. Evolutionary features include heterogeneity, colonel diversity, clonal selection process (in space), and

Table 2 | An initial classification scheme

Type	Icon	Evo-index	Eco-index	Description
1		D1Δ1	H1R1	Like a desert, these tumours have few resources and little diversity. With low turnover, they are evolutionarily inert.
2		D1Δ1	H1R2	Much like normal tissue, these tumours have sufficient resources but evolve very slowly.
3		D1Δ1	H2R1	These tumours may have the best prognosis, with an immune response that probably helps to control the tumour, restricted resources and little capacity to evolve.
4		D1Δ1	H2R2	These tumours have ample resources but have also stimulated an antitumour immune response. However, they are otherwise evolutionarily inert.
5		D1Δ2	H1R1	These tumours are genetically homogeneous but are changing over time, perhaps through population bottlenecks or selective sweeps that re-homogenize the tumour.
6		D1Δ2	H1R2	These tumours are changing over time, potentially through homogenizing selective sweeps of new clones. While they may grow rapidly, with ample resources, their genetic homogeneity may make them vulnerable to therapy.
7		D1Δ2	H2R1	Predation by the immune system in these tumours may reduce genetic heterogeneity through selection against neo-antigens.
8		D1Δ2	H2R2	Natural selection may be driving the changes in these tumours and homogenizing them.
9		D2Δ1	H1R1	These tumours may be the result of the slow accumulation of clones over a long period of time or from exposure to mutagens.
10		D2Δ1	H1R2	Like a garden, these tumours support a variety of clones, are well fed and are protected from hazards such as predation, but they change little over time.
11		D2Δ1	H2R1	Accumulation of many mutations may have led to an immune response in these tumours, but they appear to be otherwise restricted in their growth and evolution.
12		D2Δ1	H2R2	These genetically diverse tumours are changing only slowly, perhaps due to a low mutation rate or relatively weak selective pressures.
13		D2Δ2	H1R1	These tumours are evolving rapidly, generating and maintaining new clones at a high rate. They are probably under selective pressure for the ability to survive and proliferate with scarce resources or otherwise escape these resource constraints.
14		D2Δ2	H1R2	With potentially the worst prognosis, these genetically diverse tumours are evolving rapidly and have plenty of resources. They should have the highest capacity to evolve in response to interventions or other changes in their environment.
15		D2Δ2	H2R1	These rapidly evolving and diverse tumours are under the dual selective pressures of resource limitations and immune predation.
16		D2Δ2	H2R2	Like a rainforest, these genetically diverse tumours are changing rapidly, with a constant churn of new clones evolving and others going extinct. Resources are abundant, although they are probably being consumed rapidly, and predation from the immune system is extensive.

D, diversity; Δ, genetic, epigenetic or phenotypic change over time; Eco-index, ecological index; Evo-index, evolutionary index; H, hazards; R, resources.

Figure 5.4 *Classification of tumor heterogeneity.* Understanding evo/eco features can help to design next-generation nanoparticles based on tumor heterogeneity.

genome instability (in time). Ecological indexes include hazards (H) such as immune cell–infiltrated chemotherapy drug and resources (R) such as oxygen, glucose, nutrients, survival signals, growth signals, space, and hypoxia. The effects of resources on the evolution of a tumor are not defined simply

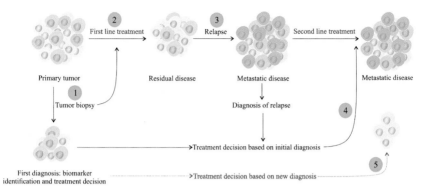

Figure 5.5 *Types of tumor heterogeneity and nanoparticle distribution.* (A) Intra-patient tumor heterogeneity: The same type of tumor in different people represents different degrees of heterogeneity, and therefore a drug cannot be prescribed for all patients with the same tumor. (B) Intertumoral heterogeneity: the primary tumor may be different from the metastatic one in terms of heterogeneity in one individual, and the drugs prescribed for the primary tumor may not work for the metastatic tumor. Primary tumors with different subclones (intratumor spatial heterogeneity) may change their heterogeneity pattern over time (intratumor temporal heterogeneity). Thus, a drug cannot be used during the whole time of the treatment course.

by their supply, depletion, and availability. Whether resources are uniform across space or heterogeneous ("patchy" or exhibiting gradients) makes a difference. Patchy resources (and hazards) create multiple habitats (for example, rich and sparse regions) that may select for different clones that can survive in those regions and may be differentially responsive to (and differentially exposed to) therapies [51]. Furthermore, patchy resources change over time, and then there is selective pressure on cells to leave regions with scarce resources and move to transient regions of more plentiful resources. Thus, ecological theory predicts that heterogeneous resources should select for invasion and metastasis. Resource gradients often lead to rapid evolution, as organisms that are able to invade more stressful environments can escape competition and flourish. The ecology of a tumor affects its evolution, and the evolution of the cells in a tumor change their ecology. Thus, before starting each anticancer regime, two items should be considered: first, the situation of eco and evo index in tumor tissue and the mapping of spatiotemporal heterogeneity in tumor tissue. The latter need several rebiopsies of primary and metastasis tumors. The prediction of interactions between this index and chemotherapy drug leads to the improved formulation and increase treatment outcome due to an increase in the remission period and delayed recurrence and metastasis rate. Not only differences at the omics levels occur as cancer progresses through time (temporal heterogeneity), but differences also arise across space (spatial heterogeneity), building colonies of tumor cells, each with a different genetic signature [52]. A complete roadmap is illustrated in Fig. 5.5.

5.1.1 Reciprocal interaction between nanoparticles and tumor ecological heterogeneity

One of the ecological consequences of tumor heterogeneity is incomplete tumor angiogenesis, which means that the high growth rate of tumor cells and the pressing need for nutrients force tumor cells to hijack other compensatory pathways, such as the co-option system. In such circumstances, in response to antiangiogenesis drugs, tumors large capillaries tend to grow laterally in the absence of active tumor neovascularization [53]. In such a case, the homogenous distribution of NPs inside the tumor can be impaired. One of the options for cancer treatments is the design of smart NPs, which are stimuli-responsive and/or multi-stage acting and in situ reacting. Meaning that these NPs are only active and operational under a specific ecologic index that is tumor hypoxia, acidity, overexpression of antigenic markers, properties of the cells, etc. [54,55]. Alternatively, NPs are engineered to modulate their microenvironment (ecology) by triggering specific chemical reactions. This way, they increase the effectiveness and response to chemotherapy or radiotherapy, improve specificity, and decrease the toxicity of nanoparticles [54].

However, none of these advanced NPs can fit the criteria for effective tumor targeting. This is because the tumor tissue is not homogeneous; thus, homogeneous distribution and localization of the nanoparticles are unlikely to occur, in particular for stimuli-responsive NPs that their activity depends upon their localization in an ecologically appropriate place (e.g., Hypoxia activated drugs (HAPs)). Plus, the stimuli-responsiveness behavior of NPs looks inefficient given that tumor features such as hypoxia and acidity are mostly gradient or patchy, and even this ecology (O2/pH gradient) is not static and may change over time. Thus, a single formulation is unlikely to be appropriate during the whole treatment course of a patient. Despite many investments, even conventional drugs (NP-free) that exploit tumor ecology, such as HAPs, have not found their way in the clinic due to the tumor heterogeneity. A similar scenario could also happen for NPs. Nanodrug may even have a different effect on metastatic tissues, and there will be no "one tablet" formulation to be used for a particular type of tumor. Misplaced localization of smart NPs in the desired region reduces NP efficacy; meanwhile, the unequal (heterogeneous) distribution may accelerate the clonal evolution and selection process and paves the way for the emergence of resistant/aggressive tumors. Another additional and mostly disregarded point is intertumor heterogeneity—that the drugs effective against primary tumors may fail to target distant metastases [56–58].

Ecological features including the degree of hypoxia, the density of blood vessels, colocalization of cancer cells with fibroblast, Blood flow and concentration of ATP, glucose, and other nutrients may have positive or negative impacts on drug release manner, localization, and retention. For example, a low density of vessels and blood flow leads to reduced NP localization. What is more alarming is that tumor stroma cells such as cancer-associated fibroblasts (CAFs) not only promote metastasis but also can lead to unspecific NP uptake and thus efficacy. Likewise, other ecological items such as hazards can negatively affect NP efficacy, in particular the type 2 macrophage. Also, nanoparticles can cause a number of changes in tumor ecology, such as nanoparticles that are in situ reactive. They may aggravate or reduce hypoxia by producing oxygen or nano-particles that cause thrombosis. Others can consume both oxygen and glucose at the same time, and even there are nanoparticles that infiltrate immune cells into tumor tissue. These NP changes affect the ecological features of the tumor, and it is unknown how it will impact the patient therapeutic outcome, knowing that going the opposite direction of what is going on inside the tumor does not always ensure delaying metastasis [59–62].

5.1.2 Interaction of nanoparticles with evolutional features

Like the ecological feature, evolutional features are also tied to nano-particles, as any changes in ecological features affect evolutional features and vice versa. The evolution of cancer clones may play an important role in their sensitivity to NPs, depending on the status of the clones and the way they are evolved [63]. Penetrations of NPs into densely populated tumors may be challenging, and dynamic evolution can promote resistance to chemotherapy. In addition, NP design can have quite a significant effect on tumor evolution, both indirectly through changes in tumor ecology and directly through drug release behavior, combination therapy, and the effect of different types of clones and even NPs on genetic and epigenetic fea-tures. Considering that each tumor has its own process for clonal evolution depending on the tumor type and ecological conditions of the tumor, which may change over time, understanding the process of colonial evolution plays an important role in the design of nanoparticles [64]. Those strategies discussed earlier with the aim of modulation of tumor features toward normalization of the tumor microenvironment may look beneficial in the first look; however, given that there are different patterns of heterogeneity due to many responses; for example, each of these normalization strategies may elicit a specific effect on each pattern, depending on the type of tumor heterogeneity and tumor diversity. That is, even starvation-based methods that reduce oxygen and nutrients and cause ischemia may force tumor cells to enter the lytic/dormant phase, while antiangiogenesis aggravates hypoxia and produce metastatic cells. Even approaches that aim to alleviate hypoxia will disturb the pattern of tumor heterogeneity as they perturb tumor population/dynamics and change the clonal evolution, thus aggravating the condition producing a therapeutic index. Even for NPs designed to produce reactive oxygen species (ROS), for example, free radicals can increase genome instability and mutation, especially in tumors that do not have significant changes in mutation and even gas therapy such as NO. All of these methods may seem effective at first glance, but due to the heterogeneous condition of the tumor and the Evo and Echo indices, they may do more harm than benefit. Characteristics of NPs that can have a deep impact on clonal evolution involves a very fast drug release behavior, such as "hit hard and fast" (conventional therapy (CT)) or very slowly (adaptive therapy (AT)) [65,66].

The strategy of "hit hard and fast" used in the design of nanodrug administers the highest possible drug dose in the shortest possible time period for eradication of the tumor. The maximum-tolerated dose (MTD) principle has been the standard of care for cancer treatment for several decades. An evolutionary flaw in this strategy is the assumption that resistant populations are not present prior to therapy. It is now clear that MTD therapy designed to kill as many cancer cells as possible, although intuitively appealing, may be evolutionarily unwise. This is because of a well-recognized Darwinian dynamic from ecology termed "competitive release," which led to the emergence of resistant cancer cells and losing tumor cell sensitivity, increasing the chance for domination of resistant cancer cells to form a large population. While studies show that the "hit hard and fast" strategy prompts more recurrence and drug resistance rate, nanoparticles with slow drug release can produce long-term remission [67]. In this line, use of the AT approach, including modulation of drug dose or holiday interval therapy, is shown to have a better response to delay tumor recurrence. As there is no "one–size–fits–all" evolutionary strategy, a specific strategy may not work for all tumor types. For tumors in which the dynamic evolution system is not very active and is homogeneous, such as lymphoma, the same high-dose method may work very well, but for tumors that are very heterogeneous, such as melanoma, the story is quite different [68].

AT strategies appear better than others for different situations. For example, with cell migration and phenotypic evolution, we can apply a strategy with less dose modulation and more emphasis on treatment vacations to keep the population sensitive. With the extreme dose changes, there is a rapid tumor response during both the growth phase and the treatment phase to either maintain the spatial structure in the case of migration or to prevent the evolution of less sensitive cells in the case of phenotypic drift. Meanwhile, using the CT strategy, the more sensitive tumors respond with complete eradication, while the less sensitive tumors eventually reoccur. In contrast, both AT schedules involving less dose modulation and interval holidays could still control the disease. While the more sensitive tumors shrank, the less sensitive ones grew, keeping the total population relatively constant [69]. Accordingly, for NP design to control dynamic tumor evolution, it is wise to use adaptive approaches than NPs with a fast release behavior to remove all drug-sensitive cells.

An additional point in the design of nanoparticles is that due to heterogeneity, tumor tissue achieves (Fig. 5.6) a clonal diversity that can act as an open source for tackling anticancer options. Another layer of complexity arises by the interaction of these clonal cells with each other and, in particular, with nontumor cells through a "tumor hijacking system." Thus, combination therapy or targeting various subclones, unlike what is believed and utilized clinically, may not always be effective in cancer therapy. The mainstream aim of research in the nanodrug delivery field is now focused on the sufficient localization of NPs at the right place/time within solid tumor tissue. However, unlike conventional chemotherapy, even if this issue is surmounted, the problem for nanodrug delivery still exists. That is, tumor cells employ several strategies to module behavior and function of resident and infiltrating cells in their own interest [70,71]. Accumulation of a large amount of smart and multifunctional NPs in the tumor may pose a big challenge for a tumor cell. It's unknown that how tumor cells can tackle this problem when it comes to NPs. However, in response to insults posed by conventional anticancer drugs, tumor cells use a "hijacking system."

(A) Universal nanoparticle formulation for cancer treatment

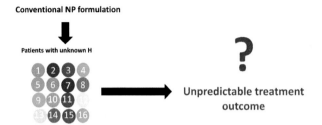

(B) Nanoparticle formulation based on Tumor heterogeneity profile

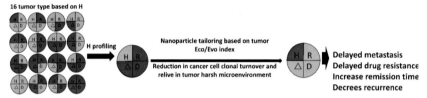

Figure 5.6 *Heterogeneous particle design to reduce the turnover of clonal selection and restoring ecological features.* As shown in the figure having a high resource, high $\triangle$, high clonal turnover, and low hazards predicts the probability for development of rapid metastasis. Thus, the treatment process should be tuned accordingly to lower D, $\triangle$ and R while increase H. Red indicates high, and blue indicates low. H, hazards (immune cell infiltration); R, resources; $\triangle$, changes over time; and D = diversity.

This way, tumor cells recruit other resident cells, molecular pathways, immune cells, organelles, and even dead cells. As in the case of immune cells, macrophage type II (TAM2) is used to promote inflammation; meanwhile, CAFs work as nonspecific trappers of NPs as well as energy sources for tumor cells in response to nutrient insults. And for NPs, as in the case for TiO_2, NPs interaction with tumor cells will spark tumor metastasis rather than tumor remedy [72]. The initial mission for multifunctional nanoparticles was to target several molecular and cellular pathways, and in advanced formats, they even can overcome drug resistance and metastasis. However, what actually occurs is that tumor cells rather neglect the cost/benefit ratio and use a hijacking strategy to fight off the harsh challenge [73]. For example, the "co-option system" allows cancer cell survival in response to the effects of a potent angiogenesis inhibitor. Thus, the hijacking system comes in handy for tumor cell survival; in particular [74], it may have the most adverse effects on multifunctional-acting nanoparticle systems (Fig. 5.7).

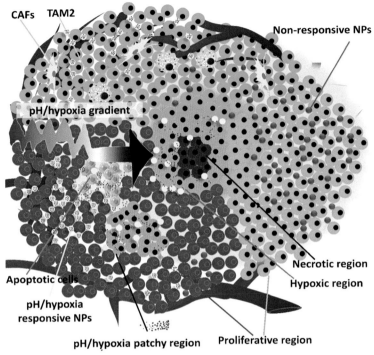

Figure 5.7 *Types of tumor heterogeneity and nanoparticle distribution.* Because the tumor is heterogeneous, the localization of the nanoparticles is certainly heterogeneous, and the two most important features for drug delivery, acidity, and hypoxia may be gradient or patch-like, in either case affecting nanoparticle activity. Also, due to the reduction of vascular density in the hypoxic regions, smart nanoparticles are not able to localize in these areas and thus fail to act responsively in nonhypoxic areas. Due to the presence of cancer-associated stroma cells, CAFs, and TAMs, the localization of nonresponsive nanoparticles (red) and smart/responsive NPs (yellow) in the heterogeneous green region is low. On the other hand, the accumulation of yellow smart NPs in the red section is heterogeneous. Drug releases occur on certain parts; as a consequence, the suppression of one subclone occurs at the expense of emerging a new subclone that increases clonal turnover in tumor tissue.

5.2 Personalized medicine in targeted delivery

Advances in technology and the development of high-tech methodologies for a better understanding of cancer, such as single-cell approaches, and the integration of different fields together, such as nanotechnology in biomedicine, have made personalized medicine indispensable for cancer therapy. Nonetheless, personalized medicine is one of the major hopes to help improve therapeutic outcomes in cancer patients by easing heterogeneous cancer-cell-derived therapeutic complications [75].

Cancer treatment becomes even more complicated with nanotechnology involvement. That is, in addition to examining the pharmacodynamics and pharmacokinetics of the chemotherapeutic drug, for example, the pharmacodynamics and pharmacokinetics of the NP should also be examined. Furthermore, all the nanodrug-delivery barriers discussed so far can differ from one patient to another and will have variable results. Currently, the main focus is on drug delivery with the aim to increase the local accumulation of NPs. And the big picture is missing that is the issue of tumor heterogeneity which highlights the role of personalized medicine even more [76]. Treatment of many diseases based on personalized medicine due to the large variety of heterogeneity (evolution and ecology) in different tumors and the role that heterogeneity plays in the pharmacodynamics and pharmacokinetics of nanoparticles pinpoints that therapeutic outcome will be different for a certain type and NP formulation with no promising results. Thus, before NP design and formulation, the tumor heterogeneity status should be evaluated to achieve the best results [77].

Given the diversity of tumor heterogeneity, even the use of avatar animals may not be helpful in the development of co-clinical trials models. Personalized medicine-based nanodrug delivery for cancer therapy calls a collective effort to achieve a common language between materialists, life scientists and clinicians, and experts in artificial intelligence. Meanwhile, cutting-edge technologies based on artificial intelligence, single-cell approaches, and microfluidics can expedite the process. By step-by-step smart formulations, one can realize NPs with maximum localization and adaptation with the heterogeneity pattern [78]. This helps to manage the plasticity of tumor heterogeneity and the clonal selection to delay recurrence and metastasis (Fig. 5.8).

With the aim of developing a patient-specific therapy, pharmacogenomics, pharmacoproteomic, and a wide variety of omic strategies have been developed in the last years [79–84]. These different techniques allow a detailed genetic and molecular profile of each patient, contributing to identify the molecular biomarkers which would affect the evolution of the disease and the response to treatments. Thus, personalized medicine is not only limited to the study of biomarkers and genetic polymorphisms [80,85] but also relies on the development of strategies for the detection of disease and the prediction of the therapeutic response. In this perspective, theranostic nanomedicines, which integrate therapeutic and imaging agents in the same nanocarrier, could contribute to developing a personalized approach in the management of grave diseases. Being conceived for

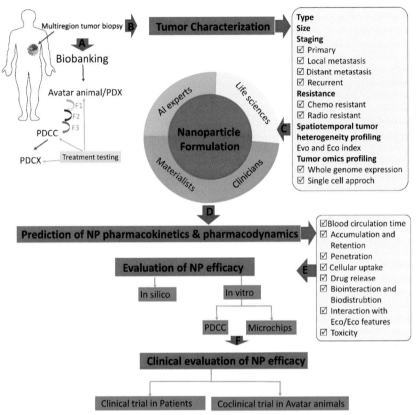

Figure 5.8 *Proposed road map for nanoparticle design based on personalized medicine.* (A) Samples are taken from different regions of the solid tumor for bio-banking. To this, slides of solid tumor are explanted into immune-comprised mice (avatar animal/patient-derived xerographs. After three passages in mice, patient-derived cell line cultures are generated and injected into mice to build patient-derived cell xerographs that can be used for drug testing. (B) In parallel, tumor biopsies are also analyzed using single-cell approaches and spatiotemporal heterogeneity imaging to profile their eco/Evo heterogeneity index. (C) The generated data is further analyzed by different experts in the field, and then (D) Further nanoparticle tailoring based on tumor characterization and the pharmacokinetic and pharmacodynamics of nanoparticles is predicted according to the properties of analyzed tumor tissues. (E) Nanoparticles are evaluated in vitro, in silico, and finally, in the case of preliminary positive results. (F) NP efficacy enters into the clinical testing.

noninvasively surveying the evolution of disease during the treatment, nanotheranostic will, in other words, drive toward the personalization of clinical treatments, which would reflect the specific characteristics of the disease in each patient. Noninvasive monitoring of drug accumulation at the target site may enable screening patients who are likely to positively respond to the treatment (characterized by high accumulation of the nanomedicine) from others who would need a different therapeutic option. Moreover, the evaluation of the accumulation also in healthy tissues would allow determining the risk of patients to develop side effects. The treatment, therefore, could be optimized in order to achieve the highest therapeutic efficiency along with the best safety profile. The possibility of having early feedbacks on the effectiveness of the treatment offers an important advantage permitting a better management of the disease, thus increasing the possibility of remission. Indeed, treatment can be tuned in real-time without waiting for traditional end points, such as tumor wrinkling. Assessing the accumulation at the target site, nanotheranostics enable the prediction of the effectiveness of treatment and may also provide a justification for the failure of the drug targeting approach in certain diseases. For example, it is well known that cancer treatment takes advantage of the passive drug accumulation in the tumor due to the previously mentioned EPR effect [86]. The efficacy of anticancer drugs such as Doxil (doxorubicin-loaded nanoparticles) and Abraxane (albumin-bound nanoparticles form of paclitaxel), which are already approved for the therapy of solid tumors (i.e., ovarian, breast cancer, and Kaposi sarcoma) [87–91], is based on this mechanism. However, the efficiency of the EPR effect is not completely understood, and individual differences are observed at the different stages of the disease with, additionally, high variability among patients. Therefore, the idea that "one fits all" and that a single therapeutic agent may be used for the treatment of all patients is not conceivable. A scarcely EPR effect seems to be the cause of the absence of response in the treatment of solid tumors such as the pancreatic adenocarcinoma [92] and the diffuse-type gastric carcinoma [93]. For these high malignant tumors, various chemotherapeutic agents have shown high efficiency in vitro, but they failed in vivo. This discrepancy is probably due to the physiology of the tumor, which includes fibrosis and hypovascularization, which oppose drug diffusion [94,95]. Moreover, significant differences can also be observed in the same tumor type, probably correlated to interpatient variability, related to the density and structural integrity of the tumor neovasculature. Thus, a rigorous evaluation of the extent of the vasculature leakage and of the drug accumulation into the tumor allows predicting the outcome of the treatment [87,96,97]. A proof of concept of this strategy has been provided by Karathanasis et al., who used iodine-labeled liposomes to predict the therapeutic response to doxorubicin-loaded liposomes treatment in a rat breast tumor model [98]. Good or bad prognosis animal groups were created by measuring the X-ray signal enhancement, which reflected the tumor accumulation of i.v.-injected iodine-labeled liposomes. After treatment, the evaluation of the tumor growth rate confirmed the previsions: a slower tumor growth rate was associated with the highest signal enhancement in the tumor and, therefore, with a leakier vasculature. This study represents a clear example of how theranostic nanomedicine could facilitate the personalized treatment of breast cancer. Indeed, clinical translation of this protocol would enable to prescreen patients, predicting which ones would have a positive outcome to the treatment due to an incomplete vasculature formation and a more important EPR effect. For potential no responder patients, another optimized and personalized option might be considered, thus avoiding the rigor of this treatment.

Preventive approaches such as mammogram screening have been adopted by a large population. Screening for BRCA1 and BRCA2 mutations also is a common practice in clinics for women in different age groups and parity status. Song et al. discussed current, and future personalized medicine approaches in breast cancer patients [99]. Because of differences in individuals' genetic backgrounds and personal susceptibility to environmental and modifiable factors, interventions do not always succeed. Increasing evidence supports personal genomic susceptibility as the major factor in responding to intervention and prevention. The approach provided by these investigators includes behavior modification for high-risk subjects (primary prevention), early detection and extensive monitoring of genetically susceptible subjects, and noninvasive treatment of early-stage cancer cases (secondary prevention), and finally, prophylactic and therapeutic intervention to slow disease progression (tertiary prevention). Based on the molecular characterization of breast cancer, individualized preventive strategies for personalized health care may be designed and implemented, although some controversies also exist, which are discussed at the end of this section. CYP2D6 (cytochrome P450, family 2, subfamily D, polypeptide 6) genotyping and its influence on breast cancer treatment by tamoxifen indicate the importance of personalized medicine in treating patients [100]. Tamoxifen is a standard treatment (endocrine therapy) for steroid receptor-positive breast cancer patients. Cytochrome P450 activates tamoxifen and forms active metabolites 4-hydroxytamoxifen and endoxifen [101]. These metabolites have two orders of magnitude of affinity toward the steroid receptor compared with tamoxifen. These compounds inhibit the proliferation of cells. CYP2D6 has different variants, and poor metabolizers and severely impaired CYP2D6 are suggested to be associated with a high recurrence of breast cancer [102]. Thus, genotyping CYP2D6 before treatment may help in predicting treatment response. An intelligent clinical decision can be made about the option of choosing strong CYP2D6 inhibitors that may inactivate active metabolites. Because pharmacogenomics-based approaches use CYP2D6 genotyping to have an idea about personal metabolizer phenotype, ethical issues must be addressed in advance. Patients and their caregivers should be well informed about the treatment strategies [103]. Raloxifene becomes an alternative choice of treatment in CYP2D6 poor metabolizer patients. Recommendations for broad CYP2D6 allele coverage and high-throughput MALDI-TOF MS/CAN (matrix-assisted laser desorption adsorption time of-flight mass spectrometry/copy number assay) have been made by Schroth et al. [104] to reduce phenotypic misclassification. Erb B2 expression-based therapy of breast cancer has shown promising results in the field of personalized medicine [105,106]. A recent report, however, indicates that routine assessment of CYP2D6 should not be used as a guide for tamoxifen treatment, and other factors should also be considered [107–109]. These investigators have suggested that aromatase inhibitors should not be administered to those patients who are pre- or premenopausal. Fleeman et al. [110] have suggested additional research on alleles other than CYP2D6 and identify patients who are responsive to treatment by tamoxifen. Norendoxifen, a metabolite of tamoxifen, is considered a potential lead compound in therapeutics due to its inhibition properties of aromatase [111]. Other reports suggest that MammPrint and Oncotype DX are current diagnostic tools that are based on expression profiling and have promising results in personalized medicine [107,111]. Future "omics" research may also add valuable information in personalized treatment of breast cancer as omics approach, including genomics, epigenomics, transcriptomics, proteomics, Metabolomics, interactomics, brings powerful ability to screen cancer cells at different stages of disease development leading to novel therapeutic target identification and validation of known targets.

A number of common treatments for colon cancer are available (chemotherapy, radiation, and surgery) [112]. Furthermore, colonoscopy screening has helped in detecting this cancer when polyps are just beginning to form. A correlation of mutations, microsatellite instability, and hypermethylation in tumors from individual patients is being completed. The information from such experiments will help to identify subgroups that are likely and not likely to respond to a particular treatment regimen [113]. This will allow patients who are likely to benefit to receive optimal care and allow those who are unlikely to benefit from avoiding unnecessary toxicity and costs. In general, when colon cancer is treated at an early stage, many patients survive at least 5 years after their diagnosis. If the colon cancer does not recur within 5 years, the disease is considered to be cured. Stage I, II, and III cancers are considered potentially curable. In most cases, stage IV cancer is not considered curable, although there are exceptions. One investigator has a different opinion about this, and according to this investigator, 5-year survival should not be considered potentially curable because late recurrences are known to arise in colon cancer and other tumor entities as well, and the 5-year survival is a rate decreasing with higher cancer stage (even in stages I–III). It has also been observed that certain therapy does not work in colorectal cancer. For example, KRAS mutations, which cover about 40% of colorectal cancers, make the tumor unresponsive to antiepidermal growth factor receptor therapy with cetuximab or panitumumab [114–116]. In terms of pharmacogenomics of colon cancer, Sarasqueta et al. [117] recently evaluated polymorphism in GSTP1, ERCC1, and ERCC2 (genes involved in the metabolism of oxaliplatin) and its correlation with the prediction of disease. In another study, Mexican patients treated with 5-fluorouracil and folinic acid predicted response to treatment with the absence or presence of polymorphism in methylenetetrahydrofolate reductase (MTHFR) gene [118]. miRNA polymorphism has been demonstrated to be associated with response to treatment with 5-fluorouracil and irinotecan [119]. If cancer started somewhere else in the body and spread to the lungs, it is called metastatic cancer to the lung. Because of the heterogeneity of cells, it is extremely difficult to treat lung cancer. Regular treatment techniques, mainly surgical and chemotherapy, have been used to treat lung cancer. Based on recent data and understanding of the genetic basis of lung cancer, EGFR, K-ras, anaplastic lymphoma kinase (ALK), MET, CBL, and COX2 are being used as therapeutic targets [107,120]. Curran recently demonstrated the utilization of crizotinib in the treatment of NSCLC. Crizotinib is an inhibitor of ALK and showed promising results. Other investigators have also observed the benefits of using crizotinib for lung cancer treatment [121,122]. Erlotinib and EGFR mutated lung cancer has also provided significant clinical results [123]. FLEX trial has also demonstrated promising results [124]. Data from histopathological examination and the patient's history also is considered in evaluating the state of the disease and its aggressiveness.

Nyberg et al. [125] studied the association between SNPs and acute interstitial lung disease in the Japanese population undergoing treatment with gefitinib. This research provided a basis for further research. In the Chinese population, ABCC1 polymorphism was found to be associated with lung cancer susceptibility in patients undergoing chemotherapy [126]. Genomic variations in EGFR and ERCC1 have also been correlated with drug response in small cell lung cancer patients [127,128]. The main screening procedures used to detect prostate cancer are the digital rectal exam and prostate-specific antigen test. Because this cancer does not cause pain and takes several years to develop, physicians and patients are faced with the challenge of identifying optimal treatment strategies for localized prostate cancer, biochemically recurrent prostate cancer, and later-stage cancer. Three treatments are very common: chemotherapy and hormonal therapy, surgery, and radiation. Age-related changes, including metastatic disease, may affect all of these therapies and shift the risk-benefit ratio of these treatments [129]. New tools, such as the Comprehensive Geriatric Assessment, are being developed to better predict who will respond to therapy. Such tools also may help in estimating the remaining life expectancy of a specific prostate cancer patient. Audet-Walsh et al. [130] demonstrated the association of several SRD5A1 (steroid 5-alpha reductase) and SRD5A2 variations as independent predictors of biochemical recurrence after radical prostatectomy in Caucasians and Asians. In another study, BCL2 polymorphism was found to be associated with adverse outcomes in prostate carcinoma patients [131].

Lymphoma is cancer in the lymphatic cells of the immune system. It is present as a solid tumor of lymphoid cells. Research is being conducted to utilize the clinical characterization of lymphoma and the integration of genomic information to identify patients who will benefit from the treatment. Lymphoma comprises mainly Hodgkin lymphoma and non-Hodgkin lymphoma, although at least 60 subtypes of lymphoma have been reported to date [132]. This cancer originates from lymph nodes but can affect other organs such as the bowel, bone, brain, and skin. Risk-stratification for all clinically identified subtypes has not been completed yet. Approaches for the stratification of lymphoma subtypes include refining clinical prognostic models for better risk stratification, use of high-throughput technology to identify biologic subtypes within pathologically similar diseases, "response-adapted" changes in therapy via imaging with [(18)F]fluoro2-deoxy-D-glucose positron emission tomography (FDG-PET), and anti-idiotype vaccines. Lymphoma treatment is accomplished by chemotherapy, radiation therapy, and bone marrow transplantation. Effective treatment for acute promyelocytic leukemia consists of identifying and developing the PML-RARA fusion gene and applying all-trans retinoic acid (ATRA) [133]. This investigation has led to the discovery of the bcr-abl fusion gene in chronic myelogenous leukemia and the development of imatinib [134]. Genetics-based drug therapy does not always work efficiently. Erlotinib and crizotinib are other genetics-based drugs with minimum efficacy in different cancers [135]. The mechanism of action of these medications is based on apoptosis. The reason for developing apoptosis-based therapies is the advantage of killing cancer cells specifically with low or minimal toxicity. These drugs were not effective because the differentiation and proliferation pathways were not affected by these drugs. In an ideal situation, the drug should inhibit all of these pathways and stop the signaling steps. To attack the final steps in the apoptosis pathway and achieve better efficacy, human recombinant DNase I-based drugs are being developed. Polymorphisms in mismatch repair genes influence response to treatment and survival in large B cell lymphoma [136]. Vagace et al. [137] identified the presence of numerous genetic variants that may have accounted for subacute methotrexate neurotoxicity in acute lymphoblastic leukemia.

5.2.1 Dual-targeted imaging platforms

Conventional single-targeted delivery systems have several limitations, including a lack of specificity for cancer cells, inability to cope with the emergence of drug resistance, and the lack of commonality between targetable receptors in different cancer types. Dual targeting strategies are promising alternatives to single-targeted delivery systems, taking advantage of two different types of cell surface receptors or TME-associated properties. Bispecific Abs that incorporate amino acid sequences that recognize two different antigen epitopes for dual targeting have been shown to enhance targeting and optimize tumor specificity. CD105 and TF are two biomarkers that are both overexpressed in pancreatic cancer. Luo et al. [138] designed a dual receptor-targeted construct consisting of a bispecific heterodimer of Fab' antibody fragments recognizing CD105 and tissue factor, using a click chemistry approach. It was dual-labeled with NIRF and PET imaging reporters (^{64}Cu-NOTA-heterodimer-ZW800) and used for the imaging of pancreatic tumors (Fig. 5.9). The PET imaging results showed higher tumor uptake in comparison with either Fab fragment

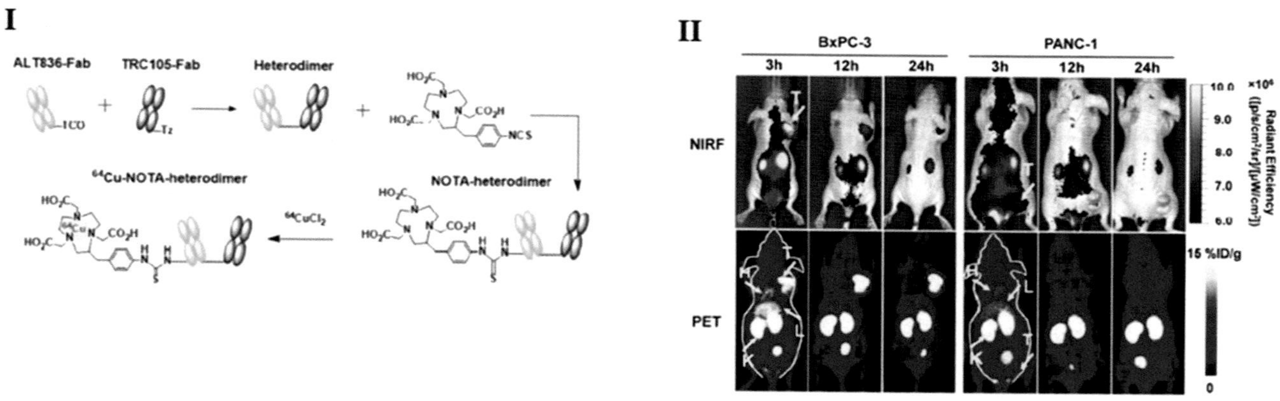

Figure 5.9 ***Dual near-infrared fluorescence and PET imaging of pancreatic tumors.*** (I) Manufacturing of ^{64}Cu-labeled heterodimer and homodimers of ALT836-F (ab') 2 and TRC105-F (ab') 2. (II) NIRF/PET imaging in mice bearing PANC-1 or BxPC-3 tumors with ^{64}Cu-NOTA heterodimer-ZW800 as a targeted imaging probe. Serial maximum intensity projections PET/NIRF images of mice bearing PANC-1 or BxPC-3 tumors at 3, 12 and 24 h following injection of targeted imaging probe. *(Reproduced with permission Luo H, et al. ImmunoPET and near-infrared fluorescence imaging of pancreatic cancer with a dual-labeled bispecific antibody fragment. Mol Pharm 2017; 14(5): 1646–1655. Copyright 2017, American Chemical Society).*

homodimer used alone. PET and NIRF imaging allowed for a clear delineation of cancer. However, the NIRF signal was significantly weaker than the PET signal [138].

Many chemotherapeutic agents enter into cells and are active only in the nucleus. Therefore, it is crucial to develop an improved delivery system to be able to target the nucleus. Surface ligand density is also an important factor that must be considered in order to achieve optimum real-time imaging of tumors. In this context, bispecific targeted imaging constructs may be a promising approach to overcome physical barriers, enhance biocompatibility, lengthen circulation time, and improve cellular uptake, for clinical diagnosis and treatment [139]. The Xiaoting Liu group [140] constructed a dual-targeted DNA tetrahedron nanocarrier loaded with DOX, with two aptamers, one to bind to MUC-1 on the cell surface, and another AS1411 to bind to nucleolin. The Dox@MUC1-Td-AS1411 construct was used for breast cancer cell imaging and drug delivery. Fluorescence imaging results showed that MUC1-Td-AS1411 could differentiate MUC1$^+$ from MUC1$^-$ cells. The DOX-loaded drug platform was effectively delivered into the nucleus, thereby killing the breast cancer cells.

Peptide targeting can also provide a modular strategy for targeting tumor tissue and molecular imaging of extracellular protease activity in vivo. For instance, activatable CPP (ACPP) is an MMP-cleavable linker that can be used in combination with cyclic-RGD binding to integrin $\alpha_v\beta_3$ for the targeting of the ECM in murine breast tumors. The cyclic-RGD-PLGC(Me)AG-ACPP loaded with chemotherapy agents allowed imaging and potent chemotherapeutic activity in mouse tumor models [141]. To achieve higher sensitivity and specificity of contrast imaging, and to overcome the poor tumor penetration of a VEGFR2 single-targeted agent, the Jing Du group [142] developed a novel dual-targeted US imaging agent using C3F8-filled PLGA NBs that were attached to dual anti-VEGFR2 and anti-HER2 mAbs. This construct could effectively penetrate the leaky tumor vasculature to target the cancer cells and lead to higher US imaging contrast compared with either of the single-targeted NBs in tumor-bearing mice. In one study Chen et al. Rationally designed a "smart" NIR fluorescence probe H2N-Cys(StBu)-Lys(Biotin)-Ser(Cy5.5)-CBT (NIR-CBT) and used it to facilely fabricate the fluorescence-quenched nanoparticle NIR-CBT-NP, whose fluorescence is turned "On" in HepG2 cancer cells for dual targeting HepG2 tumor imaging. NIR-CBT was designed to contain three components: (1) a 2-cyano-6-aminobenzothiazole (CBT) motif and a latent cysteine (Cys) group for CBT-Cys condensation 36 and subsequent self-assembly to form NIR-CBT-NP; (2) a biotin group to target biotin receptor-overexpressing tumor cells; (3) a Cy5.5 fluorophore linked to the side chain of serine (Ser) residue via an ester bond to provide the NIR fluorescence and CES cleavage site. Once the disulfide bond of NIR-CBT is reduced by a reducing agent (e.g., tris (2-carboxyethyl) phosphine, TCEP), a CBT-Cys click condensation reaction is initiated to self-assemble NIR-CBT-NP, accompanied by self-quenching of the Cy5.5 NIR fluorescence. This surface-biotinylated NIR-CBT-NP can specifically target biotin receptor-overexpressing tumor cells. After NIR-CBT-NP translocating the tumor cells, the ester bonds on their side chains are cleaved by intracellular abundant CES, turning the NIR fluorescence "On." Since NIR-CBT-NP is designed to target biotin receptor and for CES cleavage, only these two biomarker-overexpressing tumors (e.g., HepG2 tumors) are specifically targeted and precisely imaged. Compared with other dual targeting nanoparticles reported 37, they NIR-CBT-NP nanoparticle was in vitro obtained after a covalent condensation reaction which carries following two merits: (1) the synthetic process is very facile; (2) two targeting warheads on the side chains of NIR-CBT are readily switched to others by not affecting the condensation reaction. Cell and animal experiments in this work showed that the dual targeted HepG2 cells (or tumors) have the strongest NIR fluorescence [143]. Banik and coworkers, reported the design and synthesis of a dual-targeted synthetic NP to perform the double duty of diagnosis and therapy in atherosclerosis treatment regime. A library of dual-targeted NPs with an encapsulated iron oxide NP, mito-magneto (MM), with a magnetic resonance imaging (MRI) contrast enhancement capability was elucidated. Relaxivity measurements revealed that there is a substantial enhancement in transverse relativities upon the encapsulation of MM inside the dual-targeted NPs, highlighting the MRI contrast-enhancing ability of these NPs. Successful in vivo imaging documenting the distribution of MM-encapsulated dual-targeted NPs in the heart and aorta in mice ensured the diagnostic potential. The presence of mannose receptor targeting ligands and the optimization of the NP composition facilitated its ability to perform therapeutic duty by targeting the macrophages at the plaque. These dual-targeted NPs with the encapsulated MM were able to show therapeutic potential and did not trigger any toxic immunogenic response [144]. Ultrasound-targeted microbubble destruction (UTMD) is an adjuvant modality for drug delivery to localize intratumoral drug release and enhance intracellular drug accumulation. The inertial acoustic cavitation of microbubbles (MBs), including bubble implosion, microstreaming, shock waves and microjets, causes sonoporation (pore forming), which greatly improves intracellular uptake of drugs at the target site [145]. Aggregation is a common issue during preparation of the nanoparticle drug delivery system due to the interaction force between particles, which may limit its penetration into solid tumors. In view of previous study that nanoparticles can be fragmented into smaller pieces under laser irradiation to promote drug release [146], Luo and coworkers combined aggregated DPs and MBs and explored whether these large-sized drug-loaded particles could be disrupted to facilitate intracellular uptake into tumor cells, assisted by US as an external force. In one study, Luo et al. designed an aggregated dual-targeted pH-sensitive doxorubicin prodrug which was conjugated with MBs via an avidin-biotin bridge to generate a DOX prodrug-MB complex (DPMC). They examined the morphological changes of DPMC before and after US destruction, then focused on validating its tumor targeting specificity and imaging ability using in vivo fluorescence and ultrasound molecular imaging analyses. In particular, the antitumor efficacy of the complex with and without US was evaluated, both in vitro and in vivo. For prodrugs with significant cytotoxicity but relatively larger sizes, the newly generated complex assisted by US represents a promising approach to decrease size and integrate tumor imaging and therapy, providing an alternative therapeutic antitumor strategy [147]. Betulinic acid (BA) is a natural antitumor agent and has biological activity against multiple human tumor cell lines with low cytotoxicity to normal cells, while the high hydrophobicity and the short half-life of this compound limit its clinical application. In one study Lu et al. designed gelatin-based dual-targeted NPs of BA are promising to solve this problem. Hydrophobic BA is loaded in cyclodextrin to increase its solubility and prolong the circulation time in vivo. The nanoscale drug delivery systems can further enhance the bioavailability and the antitumor effect of BA and are passively targeted to the tumor tissue sites by enhanced permeability and retention effect. The RGD sequence of gelatin specifically recognizes tumor cells and brings agents into tumor cells. The nanoparticles were characterized by transmission electron microscopy, Fourier transform infrared, nuclear magnetic resonance, etc. In addition, they observed antitumor activity of the nanoparticles using both cell-based assays and mouse xenograft tumors, which proved that betulinic acid/gelatin-γ-cyclodextrin nanoparticles had a better tumor inhibition effect than betulinic acid/γ-cyclodextrin inclusion compound [148]. Amalgamation of the

ROS-responsive stimulus with nanoparticles has gained considerable interest owing to their high tumor specificity. Hypoxia plays a pivotal role in the acceleration of intracellular ROS production. Banstola et al. reported the construction of a cancer cell (PD-L1) - and ROS-responsive, dual-targeted, temozolomide (TMZ)-laden nanosystem which offers a better anticancer effect in a hypoxic tumor microenvironment. A dual-targeted system boosted permeation in the cancer cells. Hypoxic conditions elevating the high ROS level accelerated the in situ release of TMZ from anti-PD-L1−TKNPs. hyperaccumulated ROS engendered from TMZ caused oxidative damage leading to mitochondria-mediated apoptosis. TMZ fabricated in the multifunctional nanosystem (anti-PD-L1−TMZ−TKNPs) provided excellent tumor accumulation and retarded tumor growth under in vivo conditions. The elevated apoptosis effect with the activation of an apoptotic marker, DNA double-strand breakage marker, and downregulation of the angiogenesis marker in the tumor tissue following treatment with anti-PD-L1−TMZ−TKNPs exerts robust anticancer effect. Collectively, the nanoconstruct offers deep tumor permeation and high drug release and broadens the application of the ROS-responsive nanosystem for a successful anticancer effect [149].

5.2.2 Cell membrane-coated imaging agents

The complexity of biological interactions and the synthetic nature of most NPs has led to relatively poor performance of some imaging platforms within body. Overcoming these barriers using a more biomimetic design, including components of the cell membrane and its derivatives, can produce nanovesicles that can be more effectively transported within the body and interact with complex biological systems. (A) Membrane-derived molecules and components including lipids, simple sugars (e.g., mannose, galactose, sialic acid), and peptides (e.g., CD47, MUC1, fibronectin-binding protein (B) HER2/neu, etc.) have been widely explored for functionalizing NPs [150] (Fig. 5.10). In order to replicate natural membrane structures, new approaches have focused on membrane-bound biomacromolecules, carbohydrate chains, and proteins [150].

One strategy to produce biocompatible and nonimmunogenic NPs, involves using a layer of cell membrane coated around a preformed NP core. The NPs could be further functionalized with tumor-homing ligands, enhancing their circulation, active targeting, and therapeutic efficacy. Red blood cells (RBC), cancer cells, stem cells, white blood cells, and platelet cells have all been used as a source for the membrane material used as the

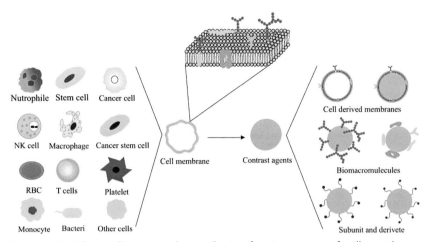

Figure 5.10 Scheme illustrating the synthesis of various types of cell membrane-coated nanoparticles.

NP coating. Each type of particle has the potential to create the next generation of nanotherapeutics and nanodiagnostics platforms [151,152]. In order to overcome the BBB, the Zhilan Chai group [153] incorporated the biotinylated form of ^{D}CDX (candoxin-derived peptide) into streptavidin/RBC membrane-coated NPs loaded with DOX. Targeting efficiency was studied in an in vitro BBB model, and in vivo studies demonstrated that the platform was capable of traversing the BBB to target brain tumors, resulting in a significant increase in the median survival of glioma-bearing mice. Because cancer cell membranes carry the full array of cancer cell membrane antigens, these antigens could be coupled to NPs and loaded with an immunological adjuvant. The resulting formulation can be used to promote a tumor-specific immune response (i.e., taken up by antigen presenting cells, APCs) for use in cancer vaccine applications [154]. Cell membrane-coated NP platforms can bridge the gap between synthetic and natural biological materials. The surface modification of imaging contrast agents with polymers could activate the immune system to different degrees. Therefore, cell membrane-coated imaging platforms could be a new approach to diagnosis and therapy. An imaging system prepared by coating UCNPs with cancer cell membranes (CCM) was shown to enhance the binding to the identical source cancer cells by flow cytometry and UCL imaging [155]. Macrophages are capable of tumor homing and can avoid reticuloendothelial system (RES) uptake, therefore cell membrane vesicles derived from macrophages (MM-vesicles) were coated onto Fe_3O_4 NPs for photothermal therapy (PTT). Fe_3O_4@MM NPs showed good biocompatibility, immune system evasion, and breast cancer targeting arising from the source macrophages [156].

One dual-modality image-guided cancer theranostic system, was described by the Yanyu Huang group [157], who developed a multifunctional smart nanosystem based on CCM vesicles (derived from leukemic cells) mixed with IDINPs and loaded with DOX-GFP-SPIO/ICG. The in vitro results showed that the CCM-camouflaged IDINPs produced ROS, induced cell death, and were "disguised" as leukemic cells, thus avoiding phagocytosis by macrophages in vivo. Furthermore, NIR laser and X-ray irradiation triggered the release of DOX from the CCM/IDINPs in GSH-enriched tumor cells with an efficient tumor-homing targeting capability in vivo. The loading of SPIO and ICG into the CCM/IDINPs enabled precise MRI and NIR imaging of the CCM/IDINPs in the tumor.

CAFs make up the majority of tumor stromal cells in the TME and are induced by several pathways operating in cancer biology. Semiconducting polymer NPs (SPNs) have been used as theranostic/imaging agents, and upon laser irradiation can generate not only NIR fluorescence and PA signals for imaging, but also singlet oxygen ($^1O_2$) and heat for combined photodynamic and photothermal therapy. In one recent study, the Jingchao Li group [158] camouflaged SPNs with fibroblast cell membranes for enhanced multimodal cancer photo theranostic. *In vivo* fluorescence and PA imaging of tumors in living mice revealed that the platform preferentially targeted CAFs, providing amplified NIR fluorescence and PA signals for tumor imaging, and enhanced the phototherapeutic efficiency of treatment. Likewise, mesenchymal stem cell (MSC) membrane-derived vesicles, with long circulation times and good tumor targeting properties, were studied to camouflage polydopamine (PDA)-coated hydrophobic Fe_3O_4 NPs, as an image-guided photothermal and siRNA delivery platform. The experimental results showed that the Fe_3O_4@PDA-siRNA@MSCs NPs displayed good MSC-mimicking ability for tumor targeting, photothermal conversion efficiency, allowed MR imaging, and also silenced the target gene in a DU145 xenograft mouse model [159]. In order to break through the physiological and pathological obstacles in the biological system and to achieve the maximum delivery of stimuli-responsive nanomedicine to tumors, Jia et al. reported a versatile nanoplatform based on pH-responsive NCs assembled from USIO NPs (USIO NCs) loaded with anticancer drug doxorubicin (DOX) and surface-coated with cancer CMs for UTMD-promoted precision tumor theranostics. In this work, citric stabilized USIO NPs were first synthesized by a solvothermal method, surface modified with ethylene diamine (EDA) to have amine groups, and then crosslinked using p-phthalaldehyde to create USIO NCs containing acidic pH-sensitive benzoic imine bonds. The formed USIO NCs were used to physically load DOX, followed by coating with cell membranes (CMs) on the surface via physical extrusion to obtain USIO NCs/DOX@CM. The resulting USIO NCs/DOX@CM were well characterized in terms of structure, morphology, composition, stability, antifouling property, pH responsive behavior, and MR relaxometry. Then, the USIO NCs/DOX@CM were used to treat B16 cells (a mouse melanoma cell line) and the xenografted tumor model in the presence or absence of UTMD. The dynamic switchable T2/T1 MR imaging of tumors was also explored. To knowledge, the design of the nanoplatform is quite unique: (1) the benzoic imine bonding used to form the USIO NCs renders them with pH-responsiveness, allowing for dynamic switchable T2/T1 MR imaging of tumors and simultaneous rapid release of DOX for tumor microenvironment-responsive precision tumor imaging and drug delivery; (2) the cancer CM-coated NCs display immune evasion-resulted improved pharmacokinetics and homologous tumor targeting specificity to the same tumors; and (3) the tumor delivery of the USIO NCs/DOX@CM can be promoted through the ultrasound-enabled sonoporation effect. Results of this study reveal that EDA-modified USIO NPs can be crosslinked with p-phthalaldehyde to form acidic-pH responsive USIO NCs that can be coated with cancer CMs and physically loaded with anticancer drug DOX. The formed USIO NCs/DOX@CM display the capabilities of immune evasion and homologous targeting due to the coating of CM, and pH responsive T_2-T_1 switchable MR imaging property and DOX release profile, thus achieving accurate dynamic T_2/T_1 MR imaging and enhanced chemotherapy of tumors in vivo. The tumor MR imaging and chemotherapy effects can be further promoted through the UTMD-rendered sonoporation effect. Overall, the prepared USIO NCs/DOX@CM may be exploited as an innovative theranostic nanosystem for elevated precision diagnosis-guided therapy of different cancer types [160].

Chimeric antigen receptors (CARs) provide T cell populations with defined antigen specificities which target tumors, regardless of the natural T cell receptor. CAR-T cells can specifically recognize tumor-associated antigen and eliminate tumor cells by ScFv which derived from monoclonal antibody heavy and light chains and expressed on the cell membrane of CAR-T cells, in a nonmajor histocompatibility complex-restricted manner. Recent successes in CAR T cell immunotherapy for CD19-positive hematological malignancies have highlighted its potential for treating solid tumors [161–163]. However, this remarkable therapeutic effect was not observed, when CAR-T cell treatment was used in solid tumors, due to certain barriers [164–166]. Aside from proper screening of tumor antigens, long-term persistence of CAR-T cells and efficient trafficking of CAR-T cells from peripheral blood circulation to tumor sites are also essential. Thus, it is conceivable to combine cell membrane coating nanotechnology with CAR-T therapy to treat solid tumors, due to the high tumor specificity of CAR-T cells and the advantage of cell membrane-camouflaged nanoparticles in drug delivery. In one study, Ma group [167] designed CAR-T membrane-coated nanoparticles were constructed for highly specific therapy for HCC. Glypican-3 (GPC3), a 580-AA heparin sulfate proteoglycan, is expressed in 75% of HCC samples, but not in healthy liver or other normal tissues [168]. Based on this fact, CAR-T cells capable of recognizing GPC3 on the cell membrane of HCC cells have been developed in recent years, which are cytotoxic to GPC3+ HCC cells [169,170]. GPC3 targeting CAR-T cells were first used to prepare CAR-T CMs. Near-infrared (NIR) dye IR780, a biodegradable photothermal and imaging agent, was then loaded in mesoporous silica nanoparticles (MSNs) to form a biodegradable core. The IR780 dye can produce fluorescence and heat under laser irradiation, and the latter effect can be employed for photothermal therapy [171,172]. Using MSNs for drug delivery is ideal due to their tunable size, good biocompatibility, lack of toxicity, tunable pore sizes (2–20 nm), and enhanced drug-loading capacity [173–175]. Taking these factors into consideration, IR780-loaded MSNs (IMs) were used in this paper for photothermal therapy to treat HCC. The IMs were coated with a layer of prefabricated CAR-T membranes using an extrusion method to fabricate tumor-specific CAR-T cell membrane-coated nanoparticles (CIMs). CIMs demonstrated enhanced tumor targeting and antitumor capabilities in vitro and in vivo. This targeted nanosystem has made significant steps toward further improving nanoparticle functionality in tumor-targeted therapy. Duan et al. designed and developed a nanoscale multimodal contrast agent with a shell layer of cRGD-decorated brain tumor cell membrane for assisting in the rapid and accurate localization of brain tumors for potential tumor margin identification as well as potential real-time image-guided resection. Taking advantage of bioorthogonal click reactions, cRGD was successfully reacted onto the surface of the brain tumor cell membrane. Such a design makes use of both synthetic cRGD ligand and natural brain tumor cell membrane derived from the tumor source, responsible for BBB penetration and homotypic targeting, to achieve an overall optimized targeting effect. Through cellular uptake and mice model experiments, cRGD CM-CPIO has been validated as a high-performance imaging contrast agent for FI, MRI, and PAI with significantly better brain tumor targeting outcomes as compared with CM-CPIO. Edging closer to a thorough understanding of brain tumor targeting mechanisms, this study provides some insights on theranostic targeting of brain tumor cells from the perspective of physio-chemistry [176]. Wang et al. proposed a new strategy to fabricate biomimetic nanocarrier with effectively enhanced BBB penetration, prolonged blood circulation, and improved tumor accumulation. Using the brain metastatic tumor CMs as camouflage, ICG-loaded polymeric nanoparticles were constructed for imaging and photothermal therapy of early brain tumors. Compared with naked NPs and NPs coated with normal CMs, the metastatic brain tumor CMs camouflaged nanoparticles exhibited a significantly higher ability to traverse both the intact BBB and BBB disrupted with different degrees. The biomimetic NPs also showed superior tumor inhibition when applied in photothermal therapy. This strategy would further promote the development of biomimetic nanoplatforms for precise diagnosis and therapy of brain tumors. It is promising for treating/diagnosing other brain-related diseases [177]. Li et al. developed a customized strategy that combines a vascular disruption agent with an antiangiogenic drug using MSNs coated with platelet membrane for self-assembled tumor-targeting accumulation. The tailor-made nanoparticles accumulate in tumor tissues through the targeted adhesion of platelet membrane surface to damaged vessel sites, resulting in significant vascular disruption and efficient antiangiogenesis in animal models. These results demonstrate the promising potential of combining vascular disruption agents and antiangiogenic drugs in a single nanoplatform for tumor eradication [178].

5.2.3 Circulating marker-based imaging
5.2.3.1 Tumor-derived extracellular vehicles
Tumor-derived extracellular vehicles (TEVs) include microvesicles, exosomes, ectosomes, and oncosomes. Exosomes are nanosized (30–150 nm) vesicles that are secreted by many cell types from both the host and the tumor into blood, urine, saliva, and ascites. They play an important role in

intercellular communication. They also reflect the phenotypic state of the parental cell, such as genetic or signaling alterations that occur in cancer cells. Exosomes are surrounded by a bilayer lipid membrane and contain many bioactive molecules such as proteins, enzymes, lipids, mRNAs, circular RNA, and microRNAs. Exosomes can pass through tissue barriers within the body and can carry out horizontal transfer of information between cancer cells. Exosomes are involved in cancer development and progression using various mechanisms, including angiogenesis, EMT, migration, metastasis, immune escape, and expansion of therapy-resistant cancer cells [179,180]. Circulating tumor-derived EVs can act as noninvasive biomarkers by measurement of their cargos, for instance, caveolin-1/S100B in melanoma, the epithelial cell adhesion molecule (EpCAM) in ovarian cancer, glypican-1 in pancreatic cancer, integrin $\alpha6\beta4$ and integrin $\alpha6\beta1$ in lung metastases, miR-17-92a in colon cancer recurrence, etc. [181]. In comparison with traditional tissue biopsies, the higher sensitivity and specificity of exosomes as tumor-specific diagnostic markers could be used for early-stage cancer diagnosis, monitoring, and prognostic evaluation. The restrictive nature of the BBB creates a major challenge for brain drug delivery, with current nanomedicines lacking the ability to cross the BBB. Extracellular vesicles have been shown to contribute to the progression of a variety of brain diseases, including metastatic brain cancer, and have been suggested as promising therapeutics and drug delivery vehicles. However, the ability of native tumor-derived extracellular vesicles to breach the BBB and the mechanisms involved in this process remain unknown. Morad et al. demonstrated that tumor-derived extracellular vesicles could breach the intact BBB in vivo, and by using state-of-the-art in vitro and in vivo models of the BBB, they have identified transcytosis as the mechanism underlying this process. Moreover, high spatiotemporal resolution microscopy demonstrated that the endothelial recycling endocytic pathway is involved in this transcellular transport. They further identify and characterize the mechanism by which tumor-derived extracellular vesicles circumvent the low physiologic rate of transcytosis in the BBB by decreasing the brain endothelial expression of rab7 and increasing the efficiency of their transport. These findings identify previously unknown mechanisms by which tumor-derived extracellular vesicles breach an intact BBB during the course of brain metastasis and can be leveraged to guide and inform the development of drug delivery approaches to deliver therapeutic cargoes across the BBB for the treatment of a variety of brain diseases including, but not limited to, brain malignancies [182].

Exosomes possess inherent biocompatibility, the ability to evade the immune system, resistance to degradation, stability in blood circulation (due to the possession of a negative zeta potential), and the ability to target particular cell types via transmembrane proteins expressed on the exosome surface. Drugs including curcumin, paclitaxel, doxorubicin, exogenous siRNAs, and antitumor miRNAs are some examples of cargos that have been delivered when encapsulated in EVs as delivery vehicles [183]. The exosomal membrane can be further modified by attaching targeting moieties to enhance tissue-specific homing and facilitate targeted drug delivery. For example, mesenchymal cell-derived exosomes were engineered to carry siRNA specific to oncogenic $KRAS^{G12D}$ (a common mutation in pancreatic cancer) and could enhance micropinocytosis by a CD47 dependent pathway, increasing the overall survival rate of mice with pancreatic cancer [184]. Bose et al. [185] constructed Cy5-antimiR-21-loaded TEVs derived from 4T1 cells used to camouflage gold-iron oxide NPs (GIONs). The multifunctional TEV-GION-NP theranostic platforms acted as a multimodal contrast agent for T2-weighted MRI in vitro. The in vivo biodistribution, tumor accumulation, and antitumor activity suggested it was promising for cancer imaging and therapy (Fig. 5.11).

Biodistribution and targeting of EVs administered via various delivery routes have been studied by in vivo tracking of the EVs to target organs. Monitoring over time has been performed both directly (e.g., lipophilic tracer dyes, radionuclides, and magnetic particles) and indirectly (e.g., transduction of a reporter gene) [186]. Specific labeling of EVs has been carried out by expression of fluorescent proteins fused with the EV membrane proteins CD63 proteins, C1C2 peptide, and luciferase mRNA [187]. The bioengineering of the parental cells and the use of extracellular vesicle mimetics (aka artificial nanovesicles) might be helpful to improve the performance of EVs and overcome the problem of the small quantities of exosomes naturally produced by cells [183]. The development of exosomal proteomics related to cancer, as well as improved microfluidic techniques for detecting and isolating exosomes, will likely improve their utility for cancer diagnosis.

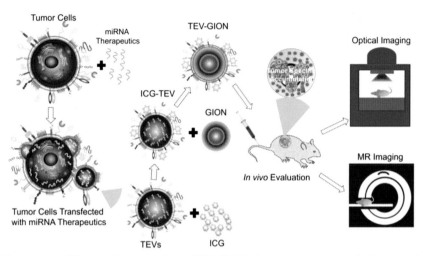

Figure 5.11 Scheme of preparation of TEV-GION-NP nanotheranostic platforms and applications for therapy and imaging. *(Reproduced with permission JC Bose R, et al. Tumor cell-derived extracellular vesicle-coated nanocarriers: an efficient theranostic platform for the cancer-specific delivery of anti-miR-21 and imaging agents. ACS Nano 2018;12(11): 10817—10832. Copyright 2018, American Chemical Society).*

On many occasions, exosomes are functionalized to increase their specificity in given tumor tissue. Modifications usually begin in the cell of origin via genetic manipulation or metabolic labeling. After being isolated, exosomes undergo modification processes that result in a nanocarrier with an exosome as a base [188—190]. Some studies are based on the direct functionalization of the exosome. For instance, exosomes from the LIM1215 CRC cell line positive for the A33 antigen (generally overexpressed in CRC cells) were loaded with DOX and exposed to carboxyl superparamagnetic iron oxide NPs coated with anti–A33 antibodies to generate a DOX-loaded functionalized nanoplatform that targets A33 positive cells. Mice-induced LIM1215 tumors showed the lowest tumor volume after treatment with functionalized exosomes (194.63 ± 13.75 mm^3) compared with those treated with DOX (477.50 ± 93.14 mm^3) and nonfunctionalized exosomes (553.88 ± 78.06 mm^3); in addition, systemic toxicity was reduced while survival increased (61, 43 and 48 days respectively), indicating great tumor targeting [191]. Human umbilical vein endothelial cells (HUVECs)–derived exosomes were functionalized by labeling the cell of origin with biotin and avidin and PTX loading. These exosomes showed high tumor tissue-targeting activity in HepG2 tumor-bearing mice even 48 h after inoculation, thus extending their half-life in the bloodstream [192]. DOX-loaded exosomes were functionalized with cancer-specific aptamers, showing different uptake mechanisms compared with free exosomes, with a predominance of clathrin-mediated endocytosis. For the target CEM cell line (acute lymphoblastic leukemia), greater cytotoxicity (up to $\sim$ 2.5 times) was observed compared with the free drug. Conversely, in the nontarget Ramos cell line (Burkitt's lymphoma), the cytotoxicity induced by functionalized exosomes was similar to DOX and was even lower ($\sim$ 1.4-fold) when the functionalized exosomes were used in higher concentrations [193]. Sonodynamic therapy was applied using exosomes derived from the 4T1 cell line and functionalized with sinoporphyrin sodium, a porphyrin sensitizer with antitumor activity and imaging properties for theragnosis showing a metastasis inhibition capacity 10 times higher compared with the free form. Furthermore, 4T1 exosomes accumulated preferentially in 4T1 tumors (2.17-fold) compared with CT26 tumors in mice [194]. Sometimes, exosomes are functionalized with magnetic NPs. For example, DOX-loaded exosomes derived from reticulocytes were functionalized with superparamagnetic magnetite colloidal nanocrystal clusters; under magnetic field exposure (MF+), Kunming mice bearing subcutaneous hepatocellular carcinoma showed higher accumulation ($\sim$ 2.12-fold) of these nanoformulations at the tumor site compared with mice not exposed to the magnetic field (MF-). In addition, the antitumor effect of these drug-loaded magnetic exosomes was enhanced by the application of an MF, obtaining tumor sizes of 1.44, 2.97, and 4.15 cm^3 for groups treated with MF+, MF−, and DOX, respectively [195]. In other cases, an entire nanoformulation is encapsulated inside the exosome. Human serum albumin (HSA)-based NPs carrying a prodrug Pt(lau)- and encapsulated in Rex exosomes derived from RAW 264.7 macrophages showed efficient breast cancer and lung metastasis targeting, with great tumor cell uptake and death [196]. Aspirin was transformed into a hydrophilic-hydrophobic nanostructure based on poloxamer 407 (a proexosome) and loaded into exosomes derived from colon and breast cancer cell lines (HT29, MDA-MB-231, respectively). The complete nanoformulations showed higher cytotoxic activity compared with the free proexosome. Moreover, they were more toxic for their cell of origin than for the other cell line, i.e., exosomes from HT29 had an IC$_{50}$ of 0.045 µM/µg protein on HT29 cells and an IC50 of 0.215 µM/µg protein on MDA-MB-231 cells [197]. Exosomes obtained from the urine of patients with gastric cancer were loaded with a functionalized nanoformulation based on gold NPs and Ce6 (a photosensitizer with apoptosis-inducing activity) for theragnosis with targeted photodynamic therapy. This nanosystem showed a long half-life in blood and a great antitumor effect with a high survival rate ($\geq$50 days) in nude mice bearing tumors induced by MGC-803 cells compared with control mice ($\leq$30 days) [198]. Piffoux et al. directly fused EVs from various cell types (HUVEC, C3H MSC, MDCK) with liposomes mediated by PEG to synthesize hybrid nanoformulations. A fluorescent antitumor photosensitizer (mTHPC) was loaded in these nanoformulations to treat cancer cells with photodynamic therapy, showing fluorescence intensity in cells three to four times higher than for free mTHPC or mTHPC-loaded liposomes [199]. Similarly, hybrid vesicles synthesized by fusing liposomes and EVs from murine macrophages (J774A.1) and loaded with DOX showed higher cytotoxic activity in breast cancer cells (4T1) [200]. Other studies focus on the de novo synthesis of exosomes based on the properties of natural exosomes derived from tumor cells but avoiding the disadvantages of these nanoplatforms, such as their oncogenic activity -whose causes have not yet been fully identified- and their limited production [201].

5.2.3.2 Circulating tumor cells and cell-free nucleic acids

Another novel and noninvasive approach to the early diagnosis of cancer is the detection of liquid biopsy-based biomarkers, such as cell-free DNA (cfDNA) or circulating tumor DNA (ctDNA). cfDNAs can be isolated from the plasma and urine of cancer patients. However, the dynamic changes (both qualitative and quantitative) in cfDNAs occurring throughout the different stages of cancer progression must be fully understood before they can be used as biomarkers for cancer and for identifying cancer relapse. The average length of cfDNA fragments found in the blood of healthy individuals and in patients diagnosed with malignant tumors is 70–200 bp and $1-200 < $ kb, respectively [202]. These high- and low molecular-weight DNA strands are likely derived from the necrotic and apoptotic cells that enter the circulation [203]. Circulating biomarkers are of great interest, especially when biopsies of the primary or metastatic tumor are not available. They could provide a longitudinal analysis method for molecular profiling of cancer cells, assessing minimal residual disease in the nonmetastatic setting, and monitoring response to systemic therapy [204].

High levels of ctDNA have been associated with poor overall survival and increased tumor burden [205,206]. ESR1 variants in cfDNA were specifically correlated with a shorter duration of endocrine treatment effectiveness in metastatic breast cancer patients [207]. Recently, PIK3CA variant detection in cfDNA was established as a companion diagnostic in clinical practice for hormone receptor-positive (HR+), human epidermal growth factor receptor 2-negative (HER2−) metastatic breast cancer (MBC) patients by FDA approval of the selective PI3Kα inhibitor Alpelisib for patients presenting PIK3CA variants in tumor tissue or plasma. cfDNA can be obtained without prior enrichment [208], and cfDNA assays have high sensitivity and reproducibility [209]. However, over the entire life span of the cfDNA, the ctDNA fraction is small [208]. The undisputable prognostic value of circulating tumor cell (CTC) enumeration in MBC was already shown 15 years ago [210] and was confirmed in large metastudies [211]. The enormous advantage of CTCs is the opportunity to analyze genomic, transcriptomic, and proteomic parameters. Regarding the mutational analysis of CTCs, the minimal number of CTCs and the consequently marginal DNA yield [212] led to the integration of whole-genome amplification before sequencing [213−217]. Interestingly, ESR1 variants were detected in CTCs and might indicate the impaired effect of aromatase inhibitor treatment in MBC patients. CTCs are viable cells actively migrating into the circulation as potential seeds of metastasis, while cfDNA is mostly generated by necrosis and apoptosis and might instead represent dying cells [218]. Importantly, the differences between cfDNA and CTCs can be regarded as a chance for comprehensive real-time disease profiling from the same patient material. Interestingly, a mutational analysis of cfDNA and transcriptional analysis of CTCs using both analyses in parallel from matched minimized blood volume revealed synergistic information [219]. Only a few comparison studies characterizing cfDNA and CTCs have been published, but cfDNA and CTCs were mostly either isolated from samples taken at different time points or from blood samples drawn into different preservative blood tubes [220−222]. For some studies, the isolation and molecular characterization of cfDNA and CTCs from matched EDTA blood samples were described, but the required blood volume was around 20 mL [223,224]. For appropriate comparability, it would be desirable to use the same blood sample with a minimized volume, drawn and stored under the same conditions for the isolation of both analyses in order to reach an unbiased, comprehensive liquid biopsy in an "all from one tube" format. ctDNAs are generally detected using microarray-based comparative genome hybridization, single-nucleotide polymorphism (SNP) analysis, massively parallel sequencing, or next-generation sequencing.

The low sensitivity and the expense of these methods are limitations for the widespread and accurate detection of cfDNAs [225]. One strategy to improve the methylation-specific PCR (MSP) technique is to use fluorescence-based (i.e., TaqMan) probes to facilitate the quantitative detection of DNA methylation without requiring further manipulation in the PCR step [226]. Similar to TaqMan probes, QDs possessing high photostability and a large dynamic range have also been used as FRET donors to detect methylated DNA. 5-amino-propargyl-2′-deoxycytidine 5′-triphosphate coupled to a Cy5 fluorescent dye served as a FRET acceptor in an assay for methylated DNA targets. The sensitivity of the Cy5-dCTP QD-FRET system was best when using multilabeled products [227].

Whole CTCs are a rare and heterogeneous population of cancer cells found in peripheral blood, which are a marker of tumor dissemination and progression. They are an attractive surrogate biomarker that could be useful in cancer diagnosis and as a prognostic indicator. The HER2 status in patients with breast cancer has been established by the analysis of CTCs [204]. The assessment of ctDNAs and CTC biomarkers is currently being incorporated into clinical trials. The in vivo monitoring of CTCs via targeted imaging modalities might provide more information about the role of these markers in tumor metastasis, mechanisms of drug resistance, and improved patient assessment.

5.2.4 Targeted cancer stem cell imaging

CSCs represent a minor subpopulation ($\sim$1% of all cancer cells) within human tumors. They possess the highest tumorigenic potential, can generate heterogeneous progeny within the bulk cancer fraction, and are thought to be responsible for much (if not most) treatment resistance. CSCs are characterized by features including the ability for self-renewal, developing into multiple lineages, and the potential to proliferate extensively [228]. CSCs are also involved in invasion and distant metastasis through the EMT/MET phenomenon. Targeting CSCs by binding to their overexpressed specific biomarkers (Table 5.1) might provide information about tumor prognosis and response in the future [228,229]. Some CSC targeting strategies are shown in Fig. 5.12.

Therapies that target CSCs in combination with conventional chemotherapy have already reached clinical trials. However, within the field of imaging and diagnosis, CSCs remain a topic of intense debate. Similar to cancer cells, in vivo imaging modalities using optical, nuclear, and magnetic resonance reporters are currently being employed to

Table 5.1 A list of known solid tumor cancer stem cell-related molecular markers for targeted therapy and diagnosis.

Breast	Colorectal	Glioma	Lung	Ovarian	Pancreatic	Prostate	Bladder
			CSC markers				
a6-integrin	ABCB5+	a6-integrin	ABCG2+	CD24−	ABCG2+	a6b1-integrin	Side
ALDH1+	ALDH1+	CD15+	ALDH1+	CD44+	ALDH1+	a6-integrin	Population
CD24−	b-catenin	CD90+	CD90+	CD117+	c-Met+	ALDH1+	
CD44+	CD24−	CD13+	CD117+	CD133+	CD24+	CD44+	
CD90+	CD26+	Nestin+	CD133+		CD44+	CD133+	
CD133+	CD29+				CD133+	CD166+	
Hedgehog-Gli	CD44+				CXCR4+	Trop2	
	CD133+				EpCAM+		
	CD166+				Nestin+		
	EpCAM+						
	LGR5+						

Targeting tumor microenvironment markers/factors

Bone microenvironment/Immune-niche/Angiogenesis/Spleen microenvironment/Lymphoid tissue/Hypoxia/Vasculogenesis

Targeting key signaling pathways

Wnt/NF-kB/Notch/PI3k/AKT/MAPK/TGF-b/mTOR/Hedgehog/Ras/Raf/Mek/Erk

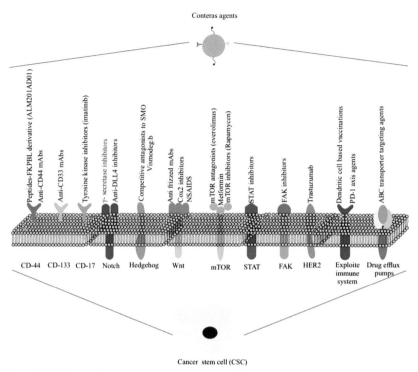

Figure 5.12 *Strategies to target cancer stem cells.* Many strategies aimed at eradicating CSCs have been developed. Targeting strategies consist of cell surface markers, immune system modulation, cell signaling pathways, and drug efflux pump inhibition to sensitize cells to imaging.

investigate the complexity underlying the behavior of CSCs. Furthermore, since CSCs are very rare in biological samples, the main concern in optical imaging is choosing a sufficiently sensitive reporter molecule and the best imaging modality. The leucine-rich repeat-containing G-protein coupled receptor 5 (LGR5) is considered to be a bona fide marker of CSCs. Researchers have used [89]Zr immunoPET to evaluate and select efficient anti-LGR5 mAbs (8F2 and 9G5) for the development of Ab-drug conjugates (ADCs), imaging, and monitoring of LGR5-positive tumor response to therapy [230]. They demonstrated that the 8F2-based ADC was more effective for toxin delivery to LGR5-positive tumors and suggested [89]Zr-labeled anti-LGR5 mAbs could be used to stratify tumors for the best response to LGR5-targeted ADC therapy. Another study evaluated the potential of a PCa-specific PC204 peptide to target CD133 and the EpCAM, two known transmembrane glycoprotein markers that are overexpressed on PCa CSCs [231]. They found that PC204 had a strong affinity for EpCAM$^+$, CD133$^+$, and CD133$^-$ CSCs in the PCa cell line and may be a promising molecular imaging platform for resistant solid tumors.

As an example of a surface marker for targeting CSC signaling pathways, the Tang group [232] targeted the ϒ-secretase enzyme, which plays an important role in the Notch signaling pathway. They used N-[N-(3, 5-fluorophenyl acetyl-L-alanyl)]-S-phenylglycine-methyl ester (DAPT) as an inhibitor of ϒ-secretase, plus HA to increase the biocompatibility and biodegradability. More specifically, HA could bind to CD44$^+$ CSCs. In vivo, MR and PA imaging results showed that the nanoprobe accumulated in the CSC microenvironment. Another approach for targeting CSCs is the use of reporter genes. The Liu group used luc2 fused to the eGFP coding sequence for designing a dual-function bioluminescence-fluorescence imaging reporter probe for breast CSCs. In vitro and in vivo results demonstrated the reporter gene was suitable for CSC targeting [233]. Proteasome activity is another candidate as a target for CSCs. Considering the fact that 26S proteasome activity is reduced in CSCs, Vlashi et al. [234] engineered cancer cells to express fluorescent fusion protein ZsGreen-ornithine decarboxylase that accumulated in cells with reduced 26S proteasome activity. The ZsGreen-positive cells could be tracked using in vivo fluorescence imaging. Results showed that the proteasome could be a suitable candidate for targeting of CSCs. The development of better techniques with higher imaging resolution and better contrast to localize CSCs will be required for their clinical detection and eradication.

Functionalization of graphene quantum dots (GQDs) with rhodamine B derivative was reported by Voelcher et al. [235]. These N-(rhodamine B)-functionalized GQDs (RBD-GQDs) have high sensitivity toward Fe^{3+} in CSCs, with a 0.02 mM detection limit and 43% quantum yield. The researchers first identified and sorted pancreatic CSCs using the CD44 surface marker, followed by incubation with different concentrations of Fe^{3+} and further incubation with RBD-GQDs. As RBD-GQDs could penetrate the membrane of pancreatic CSCs, strong intracellular fluorescence was observed at 580 nm. Fluorescence-activated cell sorting and magnetic-activated cell sorting (MACS) are common methods for isolating CSCs using an antibody against a cell surface marker. However, Sun et al. developed the metabolic labeling of CSCs with azido sugars to isolate CSCs, and they have also simultaneously performed imaging of CSCs [236]. It should be noted that cancer has a dynamic metabolic process, while CSCs have a quiescent slow-cycling phenotype. They first added Ac4ManNAz (50 mM) to both CSCs and cancer cells for 72 h, until it was incorporated into all cells, followed by the addition of Ac4ManNAc (800 mM) and 48 h incubation to allow competition with Ac4ManNAz during the metabolic process. Due to the static nature of CSCs, Ac4ManNAz remains on the surface of CSCs, which facilitates the later interaction with added DBCO-Biotin during click chemistry. Live colon CSCs from HCT-116 were isolated by simple interaction with avidin-conjugated magnetic beads. By substituting Dylight-488 avidin bound to magnetic beads, live colon CSC imaging was performed. After isolation by MACS, the group established two typical assays for the validation of CSCs: tumor spheroid formation efficiency assay (SFE) and colony formation assay (CFE). The Dynabead-captured cells showed significantly higher SFE and CFE than flow-through cells. Also, AGIS (AgGaxIn(1x)S2) QD-conjugated low-density lipoprotein (LDL) nanocomposites (NC) were constructed by Zhou et al. as the first 3D imaging material for colon CSCs (CCSCs) using nonspecific markers [237]. The LDL-based nanocomposite was fabricated with native human LDL, and the surface was modified with cationic polyelectrolyte (PDDA), followed by the coupling of negatively charged fluorescent QDs via electrostatic interaction. Targeted imaging was possible as LDL receptors are overexpressed in CSCs, and it was reported that this was the first group to elucidate the endocytic process for LDL in CSCs. The apolipoprotein ApoB-100 in LDL can be recognized by LDL receptors on the cell membrane surface. The aforementioned QD, AGIS, has the advantages of being nontoxic and possessing tunable photo-luminescence wavelengths and a high quantum yield (QY). Its character-istics were determined according to the element ratio, especially the QY was the highest (37%) when x meets a value of around 0.4 in the general formula AgGaxIn(1x)S2, with an emission wavelength of 588 nm and an average size for LDL-QDs NCs of 35.1 nm. The uptake of LDL-QDs NCs was studied by using CSCs sorted from SW480 cells with CD133 mono-clonal antibody as a marker and compared with normal colonic epithelial cells (NCM-460). The results showed that LDL-QDs efficiently accumu-lated in the CSCs, whereas fluorescence in NCM-460 cells was very weak. The uptake mechanism for LDL-QDs into CSCs was verified via a competitive inhibition assay. In order to visualize the spheroid CSCs, a 3D frame was constructed using confocal laser scanning microscopy. Aptamers are synthetic peptides or single-stranded oligonucleotides that can be chosen as a target for a specific molecule. Duan et al. first developed an

RNA aptamer against EpCAM with diagnostic and therapeutic potential [238]. An anti CD44 aptamer conjugated liposome, Apt1-Lip, was reported by Fattal et al. that selectively targets CD44-expressing tumor cells, as CD44 is a major surface marker for CSCs [239]. The group has previously reported Apt1, which targets the human isoform of CD44, and modified it with 20-F-pyrimidine to increase its stability. Aptamer functionalized-liposome (Apt1-Lip) and rhodamine fluorophore-conjugated Apt1-Lip Rhod were also used to visualize CD44 overexpressing cells. Both the CD44 positive cells (MDA-MB-231 and A549) showed high mean fluorescence intensity after treatment with Apt1-Lip-Rhod, as compared with Lip-Rhod treated groups. Moreover, cytosolic uptake of Apt1-Lip-Rhod was much more efficient on CD44+ cells (MDA-MB-231 and A549; lung cancer cells) as compared with CD44 cells (NIH/3T3; normal mouse fibroblast).

5.3 Perspectives and conclusion

Selective tumor targeting and effective delivery systems utilizing NSs have resulted in the development of novel targeting methods. Targeted nano-delivery systems are able to reach, detect, and treat various types of tumors. Many types of targets and targeting agents overlap among different human tumors. Thus, the exploration of novel molecular targets enables us to improve delivery to tumors with decreased off-target activity and less toxicity. With the aim of categorizing cancer targeting based on tumor biology, we have divided targeting strategies into five subsets including, passive targeting, TME targeting, endothelial cell targeting, general cancer cell targeting, and specific cancer cell targeting. We have also discussed new techniques and methods used for more precise cancer targeting. Despite many recent advancements in targeted delivery, there is still a long way to go, and there are many problems to overcome. These include targeted delivery structures that may still possess toxicity, the targeting moieties may not be specific, resistance or relapse is often observed in patients treated with targeted nanosystems, and most tumors currently cannot be targeted with the presently established targeted delivery systems.

Although passive and active targeting improves the accumulation and cellular uptake of NCs in tumor sites and cancer cells, even small differences in the NC size have an impact on cellular uptake and localization. In order to establish links between the nanosized particles and the targeting mechanism, the NCs need to be the ideal size to be transported out of the vasculature, penetrate into the tumors, and localize to the intended cellular compartment. On the other hand, the amount of targeted agent that can be delivered is largely independent of the percentage of the administered dose and is dependent on the precision of the targeting and the balance between passive and active targeting. In addition, an insignificant volume of administered NCs actually interacts with the cancer cells (less than 14 out of 1 million NPs injected intravenously), indicating that the majority of intratumoral NCs are either trapped in the ECM or taken up by TAMs. These off-target delivery limitations demand the re-evaluation of current targeting strategies using more quantitative approaches. Going forward, the consideration of the cancer type, subtype, and stage are critical steps in the diagnostic process by: (i) helping the clinician develop a treatment plan; (ii) giving an indication of prognosis; (iii) aiding the evaluation of the results of treatment; (iv) facilitating the exchange of information between treatment centers; and (v) contributing to further investigation regarding human cancer. Researchers need to focus on carrier-dependent targeting, combination targeting, protocols for patient selection, and routes to enable rapid and efficient clinical translation.

References

[1] Zhao Z, et al. Targeting strategies for tissue-specific drug delivery. Cell 2020;181(1):151−67.
[2] Walkey CD, et al. Protein corona fingerprinting predicts the cellular interaction of gold and silver nanoparticles. ACS Nano 2014;8(3):2439−55.
[3] Monopoli MP, et al. Physical− chemical aspects of protein corona: relevance to in vitro and in vivo biological impacts of nanoparticles. J Am Chem Soc 2011;133(8):2525−34.
[4] Monopoli MP, et al. Biomolecular coronas provide the biological identity of nano-sized materials. Nat Nanotechnol 2012;7(12):779−86.
[5] Salvati A, et al. Transferrin-functionalized nanoparticles lose their targeting capabilities when a biomolecule corona adsorbs on the surface. Nat Nanotechnol 2013;8(2):137−43.
[6] De Jong WH, et al. Particle size-dependent organ distribution of gold nanoparticles after intravenous administration. Biomaterials 2008;29(12):1912−9.
[7] Schipper ML, et al. Particle size, surface coating, and PEGylation influence the biodistribution of quantum dots in living mice. Small 2009;5(1):126−34.
[8] Sadauskas E, et al. Protracted elimination of gold nanoparticles from mouse liver. Nanomed Nanotechnol Biol Med 2009;5(2):162−9.
[9] Tsoi KM, et al. Mechanism of hard-nanomaterial clearance by the liver. Nat Mater 2016;15(11):1212−21.
[10] Liliemark E, et al. Targeting of teniposide to the mononuclear phagocytic system (MPS) by incorporation in liposomes and submicron lipid particles; an autoradiographic study in mice. Leuk Lymphoma 1995;18(1−2):113−8.

[11] Sun X, et al. Improved tumor uptake by optimizing liposome based RES blockade strategy. Theranostics 2017;7(2):319.

[12] Kwon YJ, et al. In vivo targeting of dendritic cells for activation of cellular immunity using vaccine carriers based on pH-responsive microparticles. Proc Natl Acad Sci USA 2005;102(51):18264−8.

[13] Choi HS, et al. Renal clearance of quantum dots. Nat Biotechnol 2007;25(10):1165−70.

[14] Du B, et al. Glomerular barrier behaves as an atomically precise bandpass filter in a sub-nanometre regime. Nat Nanotechnol 2017;12(11):1096.

[15] Saraiva C, et al. Nanoparticle-mediated brain drug delivery: overcoming blood−brain barrier to treat neurodegenerative diseases. J Contr Release 2016;235:34−47.

[16] Cox TR, Erler JT. Remodeling and homeostasis of the extracellular matrix: implications for fibrotic diseases and cancer. Dis Model Mech 2011;4(2):165−78.

[17] Netti PA, et al. Role of extracellular matrix assembly in interstitial transport in solid tumors. Canc Res 2000;60(9):2497−503.

[18] Kim H-Y, et al. Quantitative imaging of tumor-associated macrophages and their response to therapy using 64Cu-labeled macrin. ACS Nano 2018;12(12):12015−29.

[19] Lunov O, et al. Differential uptake of functionalized polystyrene nanoparticles by human macrophages and a monocytic cell line. ACS Nano 2011;5(3):1657−69.

[20] Dos Santos T, et al. Effects of transport inhibitors on the cellular uptake of carboxylated polystyrene nanoparticles in different cell lines. PLoS One 2011;6(9):e24438.

[21] Meng H, et al. Aspect ratio determines the quantity of mesoporous silica nanoparticle uptake by a small GTPase-dependent macropinocytosis mechanism. ACS Nano 2011;5(6):4434−47.

[22] Pack DW, Putnam D, Langer R. Design of imidazole-containing endosomolytic biopolymers for gene delivery. Biotechnol Bioeng 2000;67(2):217−23.

[23] Hu Y, et al. Cytosolic delivery of membrane-impermeable molecules in dendritic cells using pH-responsive core− shell nanoparticles. Nano Lett 2007;7(10):3056−64.

[24] Pan L, et al. Nuclear-targeted drug delivery of TAT peptide-conjugated monodisperse mesoporous silica nanoparticles. J Am Chem Soc 2012;134(13):5722−5.

[25] Nakielny S, Dreyfuss G. Transport of proteins and RNAs in and out of the nucleus. Cell 1999;99(7):677−90.

[26] Poon W, et al. A framework for designing delivery systems. Nat Nanotechnol 2020;15(10):819−29.

[27] Dai Q, et al. Quantifying the ligand-coated nanoparticle delivery to cancer cells in solid tumors. ACS Nano 2018;12(8):8423−35.

[28] Garbuzenko OB, et al. Inhibition of lung tumor growth by complex pulmonary delivery of drugs with oligonucleotides as suppressors of cellular resistance. Proc Natl Acad Sci USA 2010;107(23):10737−42.

[29] Lu M, et al. FDA report: ferumoxytol for intravenous iron therapy in adult patients with chronic kidney disease. Am J Hematol 2010;85(5):315−9.

[30] Dai X, et al. Regulation of cell uptake and cytotoxicity by nanoparticle core under the controlled shape, size, and surface chemistries. ACS Nano 2019;14(1):289−302.

[31] Oh N, Park J-H. Surface chemistry of gold nanoparticles mediates their exocytosis in macrophages. ACS Nano 2014;8(6):6232−41.

[32] Wang J, et al. Quantitative study of the interaction of multivalent ligand-modified nanoparticles with breast cancer cells with tunable receptor density. ACS Nano 2020;14(1):372−83.

[33] Poon W, et al. Elimination pathways of nanoparticles. ACS Nano 2019;13(5):5785−98.

[34] Kingston BR, et al. Assessing micrometastases as a target for nanoparticles using 3D microscopy and machine learning. Proc Natl Acad Sci USA 2019;116(30):14937−46.

[35] Stirland DL, et al. Analyzing spatiotemporal distribution of uniquely fluorescent nanoparticles in xenograft tumors. J Contr Release 2016;227:38−44.

[36] Ekdawi SN, et al. Spatial and temporal mapping of heterogeneity in liposome uptake and microvascular distribution in an orthotopic tumor xenograft model. J Contr Release 2015;207:101−11.

[37] Cuccarese MF, et al. Heterogeneity of macrophage infiltration and therapeutic response in lung carcinoma revealed by 3D organ imaging. Nat Commun 2017;8(1):1−10.

[38] Kai MP, et al. Tumor presence induces global immune changes and enhances nanoparticle clearance. ACS Nano 2016;10(1):861−70.

[39] Wu H, et al. Population pharmacokinetics of pegylated liposomal CKD-602 (S-CKD602) in patients with advanced malignancies. J Clin Pharmacol 2012;52(2):180−94.

[40] Sykes EA, et al. Tailoring nanoparticle designs to target cancer based on tumor pathophysiology. Proc Natl Acad Sci USA 2016;113(9):E1142−51.

[41] Lazarovits J, et al. Supervised learning and mass spectrometry predicts the in vivo fate of nanomaterials. ACS Nano 2019;13(7):8023−34.

[42] Ban Z, et al. Machine learning predicts the functional composition of the protein corona and the cellular recognition of nanoparticles. Proc Natl Acad Sci USA 2020;117(19):10492−9.

[43] Fourches D, et al. Quantitative nanostructure− activity relationship modeling. ACS Nano 2010;4(10):5703−12.

[44] Puzyn T, et al. Using nano-QSAR to predict the cytotoxicity of metal oxide nanoparticles. Nat Nanotechnol 2011;6(3):175−8.

[45] Paunovska K, et al. Using large datasets to understand nanotechnology. Adv Mater 2019;31(43):1902798.

[46] Yamankurt G, et al. Exploration of the nanomedicine-design space with high-throughput screening and machine learning. In: Spherical nucleic acids. Jenny Stanford Publishing; 2020. p. 1687−716.

[47] Liu R, et al. Prediction of nanoparticles-cell association based on corona proteins and physicochemical properties. Nanoscale 2015;7(21):9664−75.

[48] Khatib S, et al. Understanding the cause and consequence of tumor heterogeneity. Trends Cancer 2020;6(4):267−71.

[49] Polyak K. Tumor heterogeneity confounds and illuminates: a case for Darwinian tumor evolution. Nat Med 2014;20(4):344−6.

[50] Chakraborty J, et al. Preliminary study of tumor heterogeneity in imaging predicts two year survival in pancreatic cancer patients. PLoS One 2017;12(12):e0188022.

[51] Quetel L, et al. Genetic alterations of malignant pleural mesothelioma: association with tumor heterogeneity and overall survival. Mol Oncol 2020;14(6):1207−23.

[52] Gerdes MJ, et al. Emerging understanding of multiscale tumor heterogeneity. Front Oncol 2014;4:366.

[53] Baghban R, et al. Tumor microenvironment complexity and therapeutic implications at a glance. Cell Commun Signal 2020;18:1−19.

[54] Song G, et al. Effects of tumor microenvironment heterogeneity on nanoparticle disposition and efficacy in breast cancer tumor models. Clin Cancer Res 2014;20(23):6083−95.

[55] Eiro N, et al. Breast cancer tumor stroma: cellular components, phenotypic heterogeneity, intercellular communication, prognostic implications and therapeutic opportunities. Cancers 2019;11(5):664.

[56] Danhier F, Feron O, Préat V. To exploit the tumor microenvironment: passive and active tumor targeting of nanocarriers for anti-cancer drug delivery. J Contr Release 2010;148(2):135—46.

[57] Hui Y, et al. Understanding the effects of nanocapsular mechanical property on passive and active tumor targeting. ACS Nano 2018;12(3):2846—57.

[58] Kunjachan S, et al. Passive versus active tumor targeting using RGD-and NGR-modified polymeric nanomedicines. Nano Lett 2014;14(2):972—81.

[59] Peng J, et al. Tumor microenvironment responsive drug-dye-peptide nanoassembly for enhanced tumor-targeting, penetration, and photo-chemo-immunotherapy. Adv Funct Mater 2019;29(19):1900004.

[60] Xie J, et al. Photo synthesis of protein-based drug-delivery nanoparticles for active tumor targeting. Biomater Sci 2013;1(12):1216—22.

[61] Wu XL, et al. Tumor-targeting peptide conjugated pH-responsive micelles as a potential drug carrier for cancer therapy. Bioconjug Chem 2010;21(2):208—13.

[62] Guo X, et al. Drug delivery: dimeric drug polymeric micelles with acid-active tumor targeting and FRET-traceable drug release (Adv. Mater. 3/2018). Adv Mater 2018;30(3):1870020.

[63] Lundqvist M, et al. The evolution of the protein corona around nanoparticles: a test study. ACS Nano 2011;5(9):7503—9.

[64] Dufort S, Sancey L, Coll J-L. Physico-chemical parameters that govern nanoparticles fate also dictate rules for their molecular evolution. Adv Drug Deliv Rev 2012;64(2):179—89.

[65] Gallaher JA, et al. Spatial heterogeneity and evolutionary dynamics modulate time to recurrence in continuous and adaptive cancer therapies. Canc Res 2018;78(8):2127—39.

[66] Brehm M, et al. Self-adapting cyclic registration for motion-compensated cone-beam CT in image-guided radiation therapy. Med Phys 2012;39(12):7603—18.

[67] Aggarwal P, et al. Nanoparticle interaction with plasma proteins as it relates to particle biodistribution, biocompatibility and therapeutic efficacy. Adv Drug Deliv Rev 2009;61(6):428—37.

[68] Ferran A. Is the strategy to treat "fast, hard and for a long-term" still actual in anti-bacterial therapy? Życie Weterynaryjne 2017;92(2):119—22.

[69] Balota DA, et al. Does expanded retrieval produce benefits over equal-interval spacing? Explorations of spacing effects in healthy aging and early stage Alzheimer's disease. Psychol Aging 2006;21(1):19.

[70] Brown MR, et al. Statistical prediction of nanoparticle delivery: from culture media to cell. Nanotechnology 2015;26(15):155101.

[71] Chiu Y-L, et al. The characteristics, cellular uptake and intracellular trafficking of nanoparticles made of hydrophobically-modified chitosan. J Contr Release 2010;146(1):152—9.

[72] Madhubala V, Pugazhendhi A, Thirunavukarasu K. Cytotoxic and immunomodulatory effects of the low concentration of titanium dioxide nanoparticles (TiO_2 NPs) on human cell lines-An in vitro study. Process Biochem 2019;86:186—95.

[73] Valiente M, et al. Serpins promote cancer cell survival and vascular co-option in brain metastasis. Cell 2014;156(5):1002—16.

[74] Frentzas S, et al. Vessel co-option mediates resistance to anti-angiogenic therapy in liver metastases. Nat Med 2016;22(11):1294—302.

[75] Jo SD, et al. Targeted nanotheranostics for future personalized medicine: recent progress in cancer therapy. Theranostics 2016;6(9):1362.

[76] Kaushik A, Jayant RD, Nair M. Advances in personalized nanotherapeutics. Springer; 2017.

[77] Radhakrishnan A, et al. Pharmacogenomic phase transition from personalized medicine to patient-centric customized delivery. Pharmacogenomics J 2020;20(1):1−18.

[78] Ho D, Wang C-HK, Chow EK-H. Nanodiamonds: the intersection of nanotechnology, drug development, and personalized medicine. Sci Adv 2015;1(7):e1500439.

[79] Sakamoto JH, et al. Enabling individualized therapy through nanotechnology. Pharmacol Res 2010;62(2):57−89.

[80] Vizirianakis IS. Nanomedicine and personalized medicine toward the application of pharmacotyping in clinical practice to improve drug-delivery outcomes. Nanomed Nanotechnol Biol Med 2011;7(1):11−7.

[81] Vizirianakis IS. Pharmaceutical education in the wake of genomic technologies for drug development and personalized medicine. Eur J Pharmaceut Sci 2002;15(3):243−50.

[82] McLeod HL, Evans WE. Pharmacogenomics: unlocking the human genome for better drug therapy. Annu Rev Pharmacol Toxicol 2001;41(1):101−21.

[83] Weinshilboum R, Wang L. Pharmacogenomics: bench to bedside. Nat Rev Drug Discov 2004;3(9):739−48.

[84] Evans WE, Relling MV. Pharmacogenomics: translating functional genomics into rational therapeutics. Science 1999;286(5439):487−91.

[85] Jain K. Role of nanobiotechnology in the development of personalized medicine. 2009.

[86] Maeda H, Matsumura Y. Tumoritropic and lymphotropic principles of macromolecular drugs. Crit Rev Ther Drug Carrier Syst 1989;6(3):193−210.

[87] Northfelt DW, et al. Pegylated-liposomal doxorubicin versus doxorubicin, bleomycin, and vincristine in the treatment of AIDS-related Kaposi's sarcoma: results of a randomized phase III clinical trial. J Clin Oncol 1998;16(7):2445−51.

[88] O'Brien M, et al. CAELYX Breast Cancer Study Group: reduced cardiotoxicity and comparable efficacy in a phase III trial of pegylated liposomal doxorubicin HCl (CAELYX/Doxil) versus conventional doxorubicin for first-line treatment of metastatic breast cancer. Ann Oncol 2004;15(3):440−9.

[89] Gradishar WJ, et al. Phase III trial of nanoparticle albumin-bound paclitaxel compared with polyethylated castor oil−based paclitaxel in women with breast cancer. J Clin Oncol 2005;23(31):7794−803.

[90] Hassan M, et al. Quantitative assessment of tumor vasculature and response to therapy in kaposi's sarcoma using functional noninvasive imaging. Technol Cancer Res Treat 2004;3(5):451−7.

[91] Emoto M, et al. The blood flow characteristics in borderline ovarian tumors based on both color Doppler ultrasound and histopathological analyses. Gynecol Oncol 1998;70(3):351−7.

[92] MacKenzie MJ. Molecular therapy in pancreatic adenocarcinoma. Lancet Oncol 2004;5(9):541−9.

[93] Fuchs CS, Mayer RJ. Gastric carcinoma. N Engl J Med 1995;333(1):32−41.

[94] Sofuni A, et al. Differential diagnosis of pancreatic tumors using ultrasound contrast imaging. J Gastroenterol 2005;40(5):518−25.

[95] Takahashi Y, et al. Significance of vessel count and vascular endothelial growth factor and its receptor (KDR) in intestinal-type gastric cancer. Clin Cancer Res 1996;2(10):1679−84.

[96] Harrington KJ, et al. Effective targeting of solid tumors in patients with locally advanced cancers by radiolabeled pegylated liposomes. Clin Cancer Res 2001;7(2):243−54.

[97] van den Hoven JM, et al. Optimizing the therapeutic index of liposomal glucocorticoids in experimental arthritis. Int J Pharm 2011;416(2):471—7.

[98] Karathanasis E, et al. Imaging nanoprobe for prediction of outcome of nanoparticle chemotherapy by using mammography. Radiology 2009;250(2):398—406.

[99] Song M, Lee KM, Kang D. Breast cancer prevention based on gene—environment interaction. Mol Carcinog 2011;50(4):280—90.

[100] de Souza JA, Olopade OI. CYP2D6 genotyping and tamoxifen: an unfinished story in the quest for personalized medicine. In: Seminars in oncology. Elsevier; 2011.

[101] Brauch H, et al. Pharmacogenomics of tamoxifen therapy. Clin Chem 2009;55(10):1770—82.

[102] Brauch H, Jordan VC. Targeting of tamoxifen to enhance antitumour action for the treatment and prevention of breast cancer: the 'personalised'approach? Eur J Cancer 2009;45(13):2274—83.

[103] Hoskins JM, Carey LA, McLeod HL. CYP2D6 and tamoxifen: DNA matters in breast cancer. Nat Rev Cancer 2009;9(8):576—86.

[104] Schroth W, et al. CYP2D6 polymorphisms as predictors of outcome in breast cancer patients treated with tamoxifen: expanded polymorphism coverage improves risk stratification. Clin Cancer Res 2010;16(17):4468—77.

[105] Hatzis C, et al. A genomic predictor of response and survival following taxane-anthracycline chemotherapy for invasive breast cancer. JAMA 2011;305(18):1873—81.

[106] Arao T, et al. What can and cannot be done using a microarray analysis? Treatment stratification and clinical applications in oncology. Biol Pharm Bull 2011;34(12):1789—93.

[107] Verma M. Personalized medicine and cancer. J Personalized Med 2012;2(1):1—14.

[108] Barginear M, et al. Increasing tamoxifen dose in breast cancer patients based on CYP2D6 genotypes and endoxifen levels: effect on active metabolite isomers and the antiestrogenic activity score. Clin Pharmacol Therap 2011;90(4):605—11.

[109] Cronin-Fenton DP, Lash TL. Clinical epidemiology and pharmacology of CYP2D6 inhibition related to breast cancer outcomes. Expet Rev Clin Pharmacol 2011;4(3):363—77.

[110] Fleeman N, et al. The clinical effectiveness and cost-effectiveness of genotyping for CYP2D6 for the management of women with breast cancer treated with tamoxifen: a systematic review. 2011.

[111] Lu WJ, et al. The tamoxifen metabolite norendoxifen is a potent and selective inhibitor of aromatase (CYP19) and a potential lead compound for novel therapeutic agents. Breast Cancer Res Treat 2012;133(1):99—109.

[112] Catenacci DV, et al. Personalized colon cancer care in 2010. In: Seminars in oncology. Elsevier; 2011.

[113] Baehner FL, et al. Genomic signatures of cancer: basis for individualized risk assessment, selective staging and therapy. J Surg Oncol 2011;103(6):563—73.

[114] Amado RG, et al. Wild-type KRAS is required for panitumumab efficacy in patients with metastatic colorectal cancer. 2008.

[115] Jimeno A, et al. KRAS mutations and sensitivity to epidermal growth factor receptor inhibitors in colorectal cancer: practical application of patient selection. J Clin Oncol 2009;27(7):1130—6.

[116] Lievre A, et al. KRAS mutations as an independent prognostic factor in patients with advanced colorectal cancer treated with cetuximab. J Clin Oncol 2008;26(3):374—9.

[117] Sarasqueta AF, et al. Pharmacogenetics of oxaliplatin as adjuvant treatment in colon carcinoma. Mol Diagn Ther 2011;15(5):277—83.

[118] Castillo-Fernández O, et al. Methylenetetrahydrofolate reductase polymorphism (677 C> T) predicts long time to progression in metastatic colon cancer treated with 5-fluorouracil and folinic acid. Arch Med Res 2010;41(6):430—5.

[119] Boni V, et al. Role of primary miRNA polymorphic variants in metastatic colon cancer patients treated with 5-fluorouracil and irinotecan. Pharmacogenomics J 2011;11(6):429—36.

[120] Qian H, et al. The efficacy and safety of crizotinib in the treatment of anaplastic lymphoma kinase-positive non-small cell lung cancer: a meta-analysis of clinical trials. BMC Cancer 2014;14(1):1—6.

[121] Shaw AT, Solomon B, Kenudson MM. Crizotinib and testing for ALK. J Natl Compr Cancer Netw 2011;9(12):1335—41.

[122] Ou S-HI. Crizotinib: a novel and first-in-class multitargeted tyrosine kinase inhibitor for the treatment of anaplastic lymphoma kinase rearranged non-small cell lung cancer and beyond. Drug Des Dev Ther 2011;5:471.

[123] Chmielecki J, et al. EGFR-mutant lung adenocarcinomas treated first-line with the novel EGFR inhibitor, XL647, can subsequently retain moderate sensitivity to erlotinib. J Thorac Oncol 2012;7(2):434—42.

[124] O'Byrne KJ, et al. Molecular biomarkers in non-small-cell lung cancer: a retrospective analysis of data from the phase 3 FLEX study. Lancet Oncol 2011;12(8):795—805.

[125] Nyberg F, et al. Interstitial lung disease in gefitinib-treated Japanese patients with non-small-cell lung cancer: genome-wide analysis of genetic data. Pharmacogenomics 2011;12(7):965—75.

[126] Yin JY, et al. ABCC1 polymorphism Arg723Gln (2168G> A) is associated with lung cancer susceptibility in a Chinese population. Clin Exp Pharmacol Physiol 2011;38(9):632—7.

[127] Wu J-Y, et al. Gefitinib therapy in patients with advanced non-small cell lung cancer with or without testing for epidermal growth factor receptor (EGFR) mutations. Medicine 2011;90(3):159—67.

[128] Osawa K. SNPs in ERCC1 and drug response to cisplatin in non-small-cell lung cancer patients. Pharmacogenomics 2011;12(4):445—7.

[129] Sajid S, et al. Individualized decision-making for older men with prostate cancer: balancing cancer control with treatment consequences across the clinical spectrum. In: Seminars in oncology. Elsevier; 2011.

[130] Audet-Walsh E, et al. SRD5A polymorphisms and biochemical failure after radical prostatectomy. Eur Urol 2011;60(6):1226—34.

[131] Bachmann HS, et al. Regulatory BCL2 promoter polymorphism (− 938C> A) is associated with adverse outcome in patients with prostate carcinoma. Int J Cancer 2011;129(10):2390—9.

[132] Ramsdale E, van Besien K, Smith SM. Personalized treatment of lymphoma: promise and reality. In: Seminars in oncology. Elsevier; 2011.

[133] Nasr R, et al. Eradication of acute promyelocytic leukemia-initiating cells through PML-RARA degradation. Nat Med 2008;14(12):1333—42.

[134] Barbany G, Höglund M, Simonsson B. Complete molecular remission in chronic myelogenous leukemia after imatinib therapy. N Engl J Med 2002;347(7):539—40.

[135] Rossi D, et al. The host genetic background of DNA repair mechanisms is an independent predictor of survival in diffuse large B-cell lymphoma. Blood 2011;117(8):2405—13.

[136] Solit D, Sawyers CL. How melanomas bypass new therapy. Nature 2010;468(7326):902—3.

[137] Vagace JM, et al. Methotrexate-induced subacute neurotoxicity in a child with acute lymphoblastic leukemia carrying genetic polymorphisms related to folate homeostasis. Am J Hematol 2011;86(1):98—101.

[138] Luo H, et al. ImmunoPET and near-infrared fluorescence imaging of pancreatic cancer with a dual-labeled bispecific antibody fragment. Mol Pharm 2017;14(5):1646—55.

[139] Dianat-Moghadam H, et al. Cancer stem cells-emanated therapy resistance: implications for liposomal drug delivery systems. J Contr Release 2018;288:62—83.

[140] Liu X, et al. A dual-targeting DNA tetrahedron nanocarrier for breast cancer cell imaging and drug delivery. Talanta 2018;179:356—63.

[141] Crisp JL, et al. Dual targeting of integrin alphavbeta3 and matrix metalloproteinase-2 for optical imaging of tumors and chemotherapeutic delivery. Molecular cancer therapeutics; 2014. molcanther. 1067.2013.

[142] Du J, et al. Preparation and imaging investigation of dual-targeted C 3 F 8-filled PLGA nanobubbles as a novel ultrasound contrast agent for breast cancer. Sci Rep 2018;8(1):3887.

[143] Chen P, et al. Carboxylesterase-cleavable biotinylated nanoparticle for tumor-dual targeted imaging. Theranostics 2019;9(24):7359.

[144] Banik B, et al. Dual-targeted synthetic nanoparticles for cardiovascular diseases. ACS Appl Mater Interf 2019;12(6):6852—62.

[145] Yan F, et al. Paclitaxel-liposome—microbubble complexes as ultrasound-triggered therapeutic drug delivery carriers. J Contr Release 2013;166(3):246—55.

[146] Zheng M, et al. Single-step assembly of DOX/ICG loaded lipid—polymer nanoparticles for highly effective chemo-photothermal combination therapy. ACS Nano 2013;7(3):2056—67.

[147] Luo W, et al. Dual-targeted and pH-sensitive doxorubicin prodrug-microbubble complex with ultrasound for tumor treatment. Theranostics 2017;7(2):452.

[148] Lu S, et al. Synthesis of gelatin-based dual-targeted nanoparticles of betulinic acid for antitumor therapy. ACS Appl Bio Mater 2020;3(6):3518—25.

[149] Banstola A, et al. Hypoxia-mediated ROS amplification triggers mitochondria-mediated apoptotic cell death via PD-L1/ROS-responsive, dual-targeted, drug-laden thioketal nanoparticles. ACS Appl Mater Interf 2021;13(19):22955—69.

[150] Fang RH, et al. Cell membrane-derived nanomaterials for biomedical applications. Biomaterials 2017;128:69—83.

[151] Hu C-MJ, et al. 'Marker-of-self' functionalization of nanoscale particles through a top-down cellular membrane coating approach. Nanoscale 2013;5(7):2664—8.

[152] Li R, et al. Cell membrane-based nanoparticles: a new biomimetic platform for tumor diagnosis and treatment. Acta Pharm Sin B 2017;8(1):14—22.

[153] Chai Z, et al. A facile approach to functionalizing cell membrane-coated nanoparticles with neurotoxin-derived peptide for brain-targeted drug delivery. J Contr Release 2017;264:102—11.

[154] Fang RH, et al. Cancer cell membrane-coated nanoparticles for anticancer vaccination and drug delivery. Nano Lett 2014;14(4):2181—8.

[155] Rao L, et al. Cancer cell membrane-coated upconversion nanoprobes for highly specific tumor imaging. Adv Mater 2016;28(18):3460—6.

[156] Meng Q-F, et al. Macrophage membrane-coated iron oxide nanoparticles for enhanced photothermal tumor therapy. Nanotechnology 2018;29(13):134004.

[157] Huang Y, et al. Bioinspired tumor-homing nanosystem for precise cancer therapy via reprogramming of tumor-associated macrophages. NPG Asia Mater 2018;10(10):1002—15.

[158] Li J, et al. Cell membrane coated semiconducting polymer nanoparticles for enhanced multimodal cancer phototheranostics. ACS Nano 2018;12(8):8520—30.

[159] Mu X, et al. siRNA delivery with stem cell membrane-coated magnetic nanoparticles for imaging-guided photothermal therapy and gene therapy. ACS Biomater Sci Eng 2018;4(11):3895—905.

[160] Jia L, et al. Ultrasound-enhanced precision tumor theranostics using cell membrane-coated and pH-responsive nanoclusters assembled from ultrasmall iron oxide nanoparticles. Nano Today 2021;36:101022.

[161] Grupp SA, et al. Chimeric antigen receptor—modified T cells for acute lymphoid leukemia. N Engl J Med 2013;368(16):1509—18.

[162] Brentjens RJ, et al. Safety and persistence of adoptively transferred autologous CD19-targeted T cells in patients with relapsed or chemotherapy refractory B-cell leukemias. Blood 2011;118(18):4817—28.

[163] Kochenderfer JN, et al. B-cell depletion and remissions of malignancy along with cytokine-associated toxicity in a clinical trial of anti-CD19 chimeric-antigen-receptor—transduced T cells. Blood 2012;119(12):2709—20.

[164] Ahmed N, et al. Human epidermal growth factor receptor 2 (HER2)—specific chimeric antigen receptor—modified T cells for the immunotherapy of HER2-positive sarcoma. J Clin Oncol 2015;33(15):1688.

[165] Katz S, et al. Regional CAR-T cell infusions for peritoneal carcinomatosis are superior to systemic delivery. Cancer Gene Ther 2016;23(5):142—8.

[166] Feng K, et al. Chimeric antigen receptor-modified T cells for the immunotherapy of patients with EGFR-expressing advanced relapsed/refractory non-small cell lung cancer. Sci China Life Sci 2016;59(5):468—79.

[167] Ma W, et al. Coating biomimetic nanoparticles with chimeric antigen receptor T cell-membrane provides high specificity for hepatocellular carcinoma photothermal therapy treatment. Theranostics 2020;10(3):1281.

[168] Dargel C, et al. T cells engineered to express a T-cell receptor specific for glypican-3 to recognize and kill hepatoma cells in vitro and in mice. Gastroenterology 2015;149(4):1042—52.

[169] Gao H, et al. Development of T cells redirected to glypican-3 for the treatment of hepatocellular carcinoma. Clin Cancer Res 2014;20(24):6418—28.

[170] Jiang Z, et al. Anti-GPC3-CAR T cells suppress the growth of tumor cells in patient-derived xenografts of hepatocellular carcinoma. Front Immunol 2017;7:690.

[171] Zhang C, et al. Sentinel lymph node mapping by a near-infrared fluorescent heptamethine dye. Biomaterials 2010;31(7):1911—7.

[172] Jiang C, et al. Hydrophobic IR780 encapsulated in biodegradable human serum albumin nanoparticles for photothermal and photodynamic therapy. Acta Biomater 2015;14:61—9.

[173] Chen Y, Chen H, Shi J. In vivo bio-safety evaluations and diagnostic/therapeutic applications of chemically designed mesoporous silica nanoparticles. Adv Mater 2013;25(23):3144—76.

[174] Shao D, et al. Bioinspired diselenide-bridged mesoporous silica nanoparticles for dual-responsive protein delivery. Adv Mater 2018;30(29):1801198.

[175] Zhai Y, et al. Preparation and application of cell membrane-camouflaged nanoparticles for cancer therapy. Theranostics 2017;7(10):2575.

[176] Duan Y, et al. Biomimetic nanocomposites cloaked with bioorthogonally labeled glioblastoma cell membrane for targeted multimodal imaging of brain tumors. Adv Funct Mater 2020;30(38):2004346.

[177] Wang C, et al. Camouflaging nanoparticles with brain metastatic tumor cell membranes: a new strategy to traverse blood—brain barrier for imaging and therapy of brain tumors. Adv Funct Mater 2020;30(14):1909369.

[178] Li B, et al. Platelet-membrane-coated nanoparticles enable vascular disrupting agent combining anti-angiogenic drug for improved tumor vessel impairment. Nano Lett 2021;21(6):2588—95.

[179] Li W, et al. Role of exosomal proteins in cancer diagnosis. Mol Cancer 2017;16(1):145.

[180] Becker A, et al. Extracellular vesicles in cancer: cell-to-cell mediators of metastasis. Canc Cell 2016;30(6):836—48.

[181] Sun Y-Z, et al. Extracellular vesicles: a new perspective in tumor therapy. BioMed Res Int 2018;2018.

[182] Morad G, et al. Tumor-derived extracellular vesicles breach the intact blood—brain barrier via transcytosis. ACS Nano 2019;13(12):13853—65.

[183] Gangadaran P, Hong CM, Ahn B-C. An update on in vivo imaging of extracellular vesicles as drug delivery vehicles. Front Pharmacol 2018;9:169.

[184] Kamerkar S, et al. Exosomes facilitate therapeutic targeting of oncogenic KRAS in pancreatic cancer. Nature 2017;546(7659):498.

[185] JC Bose R, et al. Tumor cell-derived extracellular vesicle-coated nanocarriers: an efficient theranostic platform for the cancer-specific delivery of anti-miR-21 and imaging agents. ACS Nano 2018;12(11):10817—32.

[186] Gangadaran P, Hong CM, Ahn B-C. Current perspectives on in vivo noninvasive tracking of extracellular vesicles with molecular imaging. BioMed Res Int 2017;2017.

[187] Hyenne V, Lefebvre O, Goetz JG. Going live with tumor exosomes and microvesicles. Cell Adhes Migr 2017;11(2):173—86.

[188] Armstrong JP, Holme MN, Stevens MM. Re-engineering extracellular vesicles as smart nanoscale therapeutics. ACS Nano 2017;11(1):69—83.

[189] Kenari AN, Cheng L, Hill AF. Methods for loading therapeutics into extracellular vesicles and generating extracellular vesicles mimetic-nanovesicles. Methods 2020;177:103—13.

[190] Wang X, et al. Cell-derived exosomes as promising carriers for drug delivery and targeted therapy. Curr Cancer Drug Targets 2018;18(4):347—54.

[191] Li Y, et al. A33 antibody-functionalized exosomes for targeted delivery of doxorubicin against colorectal cancer. Nanomed Nanotechnol Biol Med 2018;14(7):1973—85.

[192] Wang J, et al. Chemically edited exosomes with dual ligand purified by microfluidic device for active targeted drug delivery to tumor cells. ACS Appl Mater Interf 2017;9(33):27441—52.

[193] Zou J, et al. Aptamer-functionalized exosomes: elucidating the cellular uptake mechanism and the potential for cancer-targeted chemotherapy. Anal Chem 2019;91(3):2425—30.

[194] Liu Y, et al. Focused ultrasound-augmented targeting delivery of nanosonosensitizers from homogenous exosomes for enhanced sonodynamic cancer therapy. Theranostics 2019;9(18):5261.

[195] Qi H, et al. Blood exosomes endowed with magnetic and targeting properties for cancer therapy. ACS Nano 2016;10(3):3323—33.

[196] Xiong F, et al. Pursuing specific chemotherapy of Orthotopic breast Cancer with lung metastasis from docking nanoparticles driven by bioinspired exosomes. Nano Lett 2019;19(5):3256—66.

[197] Tran PH, et al. Development of a nanoamorphous exosomal delivery system as an effective biological platform for improved encapsulation of hydrophobic drugs. Int J Pharm 2019;566:697—707.

[198] Pan S, et al. Passion fruit-like exosome-PMA/Au-BSA@ Ce6 nanovehicles for real-time fluorescence imaging and enhanced targeted photodynamic therapy with deep penetration and superior retention behavior in tumor. Biomaterials 2020;230:119606.

[199] Piffoux M, et al. Modification of extracellular vesicles by fusion with liposomes for the design of personalized biogenic drug delivery systems. ACS Nano 2018;12(7):6830—42.

[200] Rayamajhi S, et al. Macrophage-derived exosome-mimetic hybrid vesicles for tumor targeted drug delivery. Acta Biomater 2019;94:482—94.

[201] Vázquez-Ríos AJ, et al. Exosome-mimetic nanoplatforms for targeted cancer drug delivery. J Nanobiotechnol 2019;17(1):1—15.

[202] Han X, Wang J, Sun Y. Circulating tumor DNA as biomarkers for cancer detection. Dev Reprod Biol 2017;15(2):59—72.

[203] Lu T, Li J. Clinical applications of urinary cell-free DNA in cancer: current insights and promising future. Am J Cancer Res 2017;7(11):2318.

[204] De Mattos-Arruda L, et al. Circulating tumour cells and cell-free DNA as tools for managing breast cancer. Nat Rev Clin Oncol 2013;10(7):377.

[205] Dawson S-J, et al. Analysis of circulating tumor DNA to monitor metastatic breast cancer. N Engl J Med 2013;368(13):1199—209.

[206] Schwarzenbach H, Hoon DS, Pantel K. Cell-free nucleic acids as biomarkers in cancer patients. Nat Rev Cancer 2011;11(6):426—37.

[207] Takeshita T, et al. Analysis of ESR1 and PIK3CA mutations in plasma cell-free DNA from ER-positive breast cancer patients. Oncotarget 2017;8(32):52142.

[208] Lianidou E, Pantel K. Liquid biopsies. Gene Chromosome Cancer 2019;58(4):219—32.

[209] Appierto V, et al. How to study and overcome tumor heterogeneity with circulating biomarkers: the breast cancer case. In: Seminars in cancer biology. Elsevier; 2017.

[210] Cristofanilli M, et al. Circulating tumor cells, disease progression, and survival in metastatic breast cancer. N Engl J Med 2004;351(8):781—91.

[211] Bidard F-C, et al. Clinical validity of circulating tumour cells in patients with metastatic breast cancer: a pooled analysis of individual patient data. Lancet Oncol 2014;15(4):406—14.

[212] Kidess-Sigal E, et al. Enumeration and targeted analysis of KRAS, BRAF and PIK3CA mutations in CTCs captured by a label-free platform: comparison to ctDNA and tissue in metastatic colorectal cancer. Oncotarget 2016;7(51):85349.

[213] De Luca F, et al. Mutational analysis of single circulating tumor cells by next generation sequencing in metastatic breast cancer. Oncotarget 2016;7(18):26107.

[214] Paoletti C, et al. Comprehensive mutation and copy number profiling in archived circulating breast cancer tumor cells documents heterogeneous resistance mechanisms. Canc Res 2018;78(4):1110—22.

[215] Paolillo C, et al. Detection of activating estrogen receptor gene (ESR1) mutations in single circulating tumor cells. Clin Cancer Res 2017;23(20):6086—93.

[216] Pestrin M, et al. Heterogeneity of PIK3CA mutational status at the single cell level in circulating tumor cells from metastatic breast cancer patients. Mol Oncol 2015;9(4):749—57.

[217] Shaw JA, et al. Mutation analysis of cell-free DNA and single circulating tumor cells in metastatic breast cancer patients with high circulating tumor cell counts. Clin Cancer Res 2017;23(1):88—96.

[218] Wan JC, et al. Liquid biopsies come of age: towards implementation of circulating tumour DNA. Nat Rev Canc 2017;17(4):223.

[219] Keup C, et al. Cell-free DNA variant sequencing using CTC-depleted blood for comprehensive liquid biopsy testing in metastatic breast cancer. Cancers 2019;11(2):238.

[220] Heidary M, et al. The dynamic range of circulating tumor DNA in metastatic breast cancer. Breast Cancer Res 2014;16(4):1—10.

[221] Yanagita M, et al. A prospective evaluation of circulating tumor cells and cell-free DNA in EGFR-mutant non—small cell lung cancer patients treated with erlotinib on a phase II trial. Clin Cancer Res 2016;22(24):6010—20.

[222] Hodara E, et al. Multiparametric liquid biopsy analysis in metastatic prostate cancer. JCI Insight 2019;4(5).

[223] Mastoraki S, et al. ESR1 methylation: a liquid biopsy—based epigenetic assay for the follow-up of patients with metastatic breast cancer receiving endocrine treatment. Clin Cancer Res 2018;24(6):1500—10.

[224] Tzanikou E, et al. PIK 3 CA hotspot mutations in circulating tumor cells and paired circulating tumor DNA in breast cancer: a direct comparison study. Mol Oncol 2019;13(12):2515—30.

[225] Ilie M, et al. Current challenges for detection of circulating tumor cells and cell-free circulating nucleic acids, and their characterization in non-small cell lung carcinoma patients. What is the best blood substrate for personalized medicine? Ann Transl Med 2014;2(11).

[226] Eads CA, et al. MethyLight: a high-throughput assay to measure DNA methylation. Nucl Acids Res 2000;28(8). e32-00.

[227] Bailey VJ, et al. Enzymatic incorporation of multiple dyes for increased sensitivity in QD-FRET sensing for DNA methylation detection. Chembiochem 2010;11(1):71—4.

[228] Batlle E, Clevers H. Cancer stem cells revisited. Nat Med 2017;23(10):1124.

[229] Takebe N, et al. Targeting Notch, Hedgehog, and Wnt pathways in cancer stem cells: clinical update. Nat Rev Clin Oncol 2015;12(8):445.

[230] Azhdarinia A, et al. Evaluation of anti-LGR5 antibodies by immunoPET for imaging colorectal tumors and development of antibody—drug conjugates. Mol Pharm 2018;15(6):2448—54.

[231] Yeh C-Y, et al. Peptide-conjugated nanoparticles for targeted imaging and therapy of prostate cancer. Biomaterials 2016;99:1—15.

[232] Tang J, et al. Dual-mode imaging-guided synergistic chemo-and magneto-hyperthermia therapy in a versatile nanoplatform to eliminate cancer stem cells. ACS Appl Mater Interf 2017;9(28):23497—507.

[233] Liu H, et al. Cancer stem cells from human breast tumors are involved in spontaneous metastases in orthotopic mouse models. Proc Natl Acad Sci USA 2010:201006732.

[234] Vlashi E, et al. In vivo imaging, tracking, and targeting of cancer stem cells. J Natl Cancer Inst 2009;101(5):350—9.

[235] Guo R, et al. Rhodamine-functionalized graphene quantum dots for detection of Fe^{3+} in cancer stem cells. ACS Appl Mater Interf 2015;7(43):23958—66.

[236] Sun L, et al. Rapid recognition and isolation of live colon cancer stem cells by using metabolic labeling of azido sugar and magnetic beads. Anal Chem 2016;88(7):3953—8.

[237] Song J, et al. Polyelectrolyte-mediated nontoxic AgGa x In1—x S2 QDs/low-density lipoprotein nanoprobe for selective 3D fluorescence imaging of cancer stem cells. ACS Appl Mater Interf 2019;11(10):9884—92.

[238] Nakamura M, et al. Survivin as a predictor of cis-diamminedichloroplatinum sensitivity in gastric cancer patients. Cancer Sci 2004;95(1):44—51.

[239] Alshaer W, et al. Functionalizing liposomes with anti-CD44 aptamer for selective targeting of cancer cells. Bioconjug Chem 2015;26(7):1307—13.

Index

Note: 'Page numbers followed by "*f*" indicate figures and "*t*" indicate tables'.

Printed in the United States
by Baker & Taylor Publisher Services